Ellen Eigemeier

Generating vegetation reference-spectra for DOAS satellite application

Ellen Eigemeier

Generating vegetation reference-spectra for DOAS satellite application

From basic principles to first applications of the new reference-spectra

Südwestdeutscher Verlag für Hochschulschriften

Imprint
Any brand names and product names mentioned in this book are subject to trademark, brand or patent protection and are trademarks or registered trademarks of their respective holders. The use of brand names, product names, common names, trade names, product descriptions etc. even without a particular marking in this work is in no way to be construed to mean that such names may be regarded as unrestricted in respect of trademark and brand protection legislation and could thus be used by anyone.

Publisher:
Südwestdeutscher Verlag für Hochschulschriften
is a trademark of
Dodo Books Indian Ocean Ltd., member of the OmniScriptum S.R.L Publishing group
str. A.Russo 15, of. 61, Chisinau-2068, Republic of Moldova Europe
Printed at: see last page
ISBN: 978-3-8381-2412-4

Zugl. / Approved by: Mainz, Johannes Gutenberg Universität, Diss., 2010

Copyright © Ellen Eigemeier
Copyright © 2012 Dodo Books Indian Ocean Ltd., member of the OmniScriptum S.R.L Publishing group

Abstract

Vegetation-cycles are of general interest for many applications. Be it for harvest-predictions, global monitoring of climate-change or as input to atmospheric models.

Common Vegetation Indices use the fact that for vegetation the difference between Red and Near Infrared reflection is higher than in any other material on Earth's surface. This gives a very high degree of confidence for vegetation-detection.

The spectrally resolving data from the GOME and SCIAMACHY satellite-instruments provide the chance to analyse finer spectral features throughout the Red and Near Infrared spectrum using Differential Optical Absorption Spectroscopy (DOAS). Although originally developed to retrieve information on atmospheric trace gases, we use it to gain information on vegetation. Another advantage is that this method automatically corrects for changes in the atmosphere. This renders the vegetation-information easily comparable over long time-spans.

The first results using previously available reference spectra were encouraging, but also indicated substantial limitations of the available reflectance spectra of vegetation. This was the motivation to create new and more suitable vegetation reference spectra within this thesis. The set of reference spectra obtained is unique in its extent and also with respect to its spectral resolution and the quality of the spectral calibration. For the first time, this allowed a comprehensive investigation of the high-frequency spectral structures of vegetation reflectance and of their dependence on the viewing geometry.

The results indicate that high-frequency reflectance from vegetation is very complex and highly variable. While this is an interesting finding in itself, it also complicates the application of the obtained reference spectra to the spectral analysis of satellite observations.

The new set of vegetation reference spectra created in this thesis opens new perspectives for research. Besides refined satellite analyses, these spectra might also be used for applications on other platforms such as aircraft. First promising studies have been presented in this thesis, but the full potential for the remote sensing of vegetation from satellite (or aircraft) could be further exploited in future studies.

Acknowledgements

This thesis would never have been accomplished without the help and kind support of many people.

I am very much indebted to my boss at Max-Planck Institute for Chemistry in Mainz, for the courage of giving me the chance to do my PhD after leaving academia for almost 10 years, working in completely unrelated fields. His patience and expertise were invaluable support.

Thank you very much to my "PhD-father" for accepting me as his PhD-student at the Institute of Geography at Johannes Gutenberg-University in Mainz and providing steadfast support.

Another strong base of support was my PhD Advisory Committee (PAC) for the International Max Planck Research School for Atmospheric Chemistry and Physics (IMPRS).

A permanent base of support were my colleagues at the satellite group.

During measurements in the botanical garden I was intensively supported by a student intern from the University of Trier.

Many thanks to the coordinator of the International Max Planck Research School for Atmospheric Chemistry and Physics and my fellow students in the IMPRS, providing scientific exchange and social support at the same time.

Last but not least I need to thank my parents. Without their love and support none of this would have been possible.

Contents

1 Introduction 7
2 Radiative Transfer 9
   2.1 Atmosphere 10
      2.1.1 Absorption 11
         2.1.1.1 Fraunhofer lines 11
         2.1.1.2 Chemical Species in the atmosphere 11
      2.1.2 Scattering 13
         2.1.2.1 Mie 13
         2.1.2.2 Rayleigh 14
         2.1.2.3 Raman scattering and Ring-effect 14
   2.2 Vegetation 15
      2.2.1 Absorption by Pigments 17
         2.2.1.1 Chlorophyll 18
         2.2.1.2 Carotenoids 21
         2.2.1.3 Flavonoids, especially Anthocyanins 22
      2.2.2 Scattering inside the leaf 24
         2.2.2.1 Build-up of a cell and the influence on scattering properties 25
         2.2.2.2 Build-up of a leaf and the influence on scattering properties 26
      2.2.3 Scattering within the canopy 28
      2.2.3 Differences in leaf-build-up and canopy 30
3 DOAS 35
   3.1 Method 35
   3.2 Application to satellite data 38
4 Satellite Instruments and Data 42
   4.1   GOME 42
   4.2   SCIAMACHY 43
   4.3   MERIS 46
      4.3.1   Globcover and Vegetation-classes 47
   4.4 HICRU 48
5 Mini-MAX-DOAS 49
   5.1   Build-up of instrument 49
   5.2   External sources of interference on measurements 51
      5.2.1   Temperature 51
      5.2.2   Light-source 52
   5.3   Spectral calibration 54
   5.4   Spectral Resolution 55
   5.5   Spectral stability 56
6 Measurement of Vegetation Reference Spectra 57
   6.1   Methods of Measurement 57
      6.1.1   Potential Sources of Interference 57
      6.1.2   Requirements for measurements without external Interference ... 57
      6.1.3   Methods for creating reference spectra 60
   6.2   Measurements over Vegetation 62
      6.2.1   Single Measurement 62

- 6.2.2 Averaging over a single Plant ... 63
- 6.2.3 Averaging over Plant-Groups ... 63
- 6.2.4 Resolution Adjustment to Satellite instrument ... 63
- 6.3 Vegetation Reference Spectra ... 64
  - 6.3.1 Initial averages for first trials ... 64
  - 6.3.2 Refined reference spectra ... 66
- 7 Initial analysis of satellite spectra using GOME ... 70
  - 7.1 Clouds ... 71
  - 7.2 Bias correction for vegetation analysis ... 74
    - 7.2.1 Sahara ... 74
    - 7.2.2 Southern ocean ... 75
  - 7.3 Summary and Conclusion ... 76
- 8 Results of the Mini-MAX-DOAS measurements ... 77
  - 8.1 Leaf with and without pigments ... 78
  - 8.2 Reflection from upper and lower side of a leaf ... 80
  - 8.3 Comparison of spruce-needles ... 82
  - 8.4 Changes over grass during vegetation-cycle ... 85
  - 8.5 Changes in pigments and absorption in autumn ... 88
  - 8.6 Lichen ... 90
  - 8.7 Influence of incidence angles ... 92
  - 8.8 Summary ... 98
- 9 Optimizing DOAS-fit for vegetation spectra ... 99
  - 9.1 Choice of parameters for the spectral analysis ... 99
  - 9.2 Reference areas ... 103
    - 9.2.1 Selected reference areas ... 105
      - 9.2.1.1 Rainforest ... 106
      - 9.2.1.2 Deciduous forest ... 108
      - 9.2.1.3 Coniferous forest / Taiga ... 109
      - 9.2.1.4 Tundra ... 111
      - 9.2.1.5 Grassland: agriculture ... 112
      - 9.2.1.6 Grassland: unmanaged ... 115
  - 9.3 Fit results for the selected reference areas ... 116
  - 9.4 Global monthly mean maps ... 118
  - 9.5 Modified global monthly mean maps ... 121
- 10) Discussion and Conclusion ... 123
- References ... 125
- Appendix A ... 131
  - Sweet Grasses ... 133
  - Grasses ... 141
  - Coniferous trees ... 145
  - Deciduous trees ... 149
  - Deciduous others ... 159
- Appendix B ... 161

1 Introduction

Vegetation-cycles are of general interest for many applications. Be it for harvest-predictions, global monitoring of climate-change or as input to atmospheric models.

Common Vegetation Indices use the fact that for vegetation the difference between Red and Near Infrared reflection is higher than in any other material on Earth's surface. This gives a very high degree of confidence for vegetation-detection.

The spectrally resolving data from the GOME and SCIAMACHY satellite-instruments provide the chance to analyse finer spectral features throughout the Red and Near Infrared spectrum. using Differential Optical Absorption Spectroscopy (DOAS). Although originally developed to retrieve information on trace gases, we use it to gain information on vegetation. Another advantage is that this method automatically corrects for changes in the atmosphere. This renders the vegetation-information easily comparable over long time-spans.

To enable vegetation monitoring with DOAS, new and more suitable vegetation reference spectra were necessary. In this thesis, new vegetation reference spectra are produced and presented. Especially the high-frequency structures of vegetation reflectance are systematically investigated. Some first applications to satellite data examine the potential for future research.

This thesis first introduces basics of radiative transfer in the atmosphere and the influences on vegetation reflectance in Chapter 2. An introduction to DOAS follows in Chapter 3.

Chapters 4 and 5 introduce the satellites, instruments and additional data-bases. The methods to produce the new vegetation reference spectra are described in Chapter 6.

Results of initial applications of vegetation reference data are shown in Chapter 7. Chapter 8 presents the created data set of new high-frequency vegetation reflectance spectra and their dependencies on life-cycle illumination and viewing angles. Chapter 9 shows the results of first applications to satellite data. These are followed by the conclusion and outlook in Chapter 10.

The species measured for the new vegetation reference spectra are documented in Appendix A. Additional measurement-results for changing illumination and viewing angle are presented in Appendix B.

2 Radiative Transfer

The light measured by the satellites considered in this thesis originates from the sun. It passes through the atmosphere, gets scattered back either by the atmosphere or from the ground and passes through the atmosphere again. Only then is it recorded by the satellite instrument. On this long path the light is influenced on many occasions. Therefore the light impacting the atmosphere has a significantly different signal compared to the light exiting the atmosphere and being registered by the satellite. These changes have different causes.

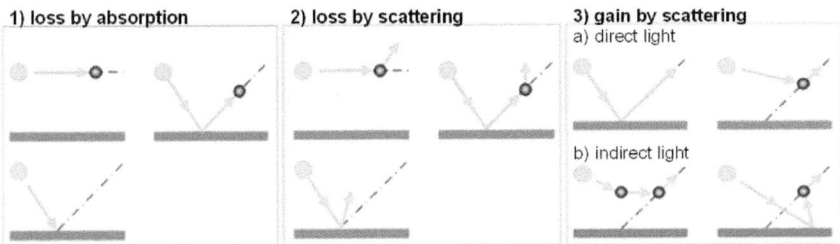

Figure 2.1
Processes causing the difference between incident sunlight and observed light at the satellite. 1) shows different absorption processes, removing light from the light-path. 2) depicts scattering processes, where the light is removed from the observed light-path into other directions. 3) depicts scenarios where light is scattered into the observed light-path, resulting in gains by scattering.
Modified from AT2-ELS

Figure 2.1 shows the most common effects on the observed light-path. In the first scenario the Figure depicts processes of light being lost on its path by absorption, either by molecules and aerosols in the atmosphere or by interaction with the ground.

The second scenario considers scattering that removes light from the satellite instrument's field of view. Scattering can happen in the atmosphere or on the ground. These processes will be discussed at a later stage in this chapter.

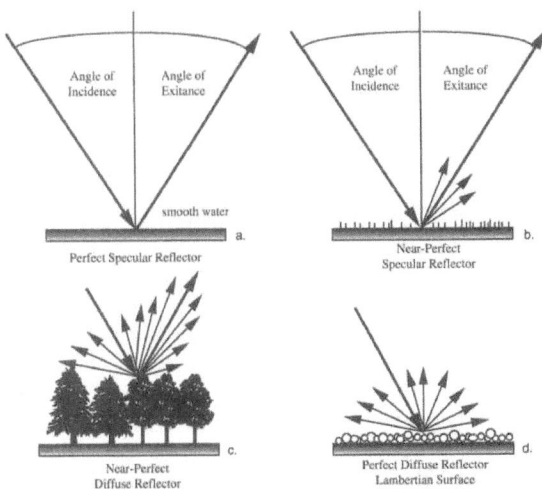

Figure 2.2
The nature of specular and diffuse reflectance. (Source: Jensen 2000)

Light is not only lost on the light path. Some processes may also produce gains to the light inside the field of view: see scenario three. This happens through light scattering from areas outside the original view of the satellite instrument into the light path reaching the satellite. This may either be direct light scattered into the path, or indirect light which had other interactions before.

Many of the processes described above consider scattering processes. However one scattering process is not the same as

9

another one, as is demonstrated in Figure 2.2. For a perfect specular reflector, the incident angle is the same as the exiting angle. But in many cases on natural surfaces this is not the case, as pointed out in the scenarios b to d. Since we are concentrating on vegetation, especially scenario c is interesting.

2.1 Atmosphere

The atmosphere is the gas layer that surrounds our planet. It encompasses all layers from those above the surface of earth to all the way out into space. The thickness of the atmosphere is not quite defined because it gets progressively thinner.

Most interesting for us is the lowest part of the atmosphere, the troposphere. Here pressure is highest and the number of molecules and aerosols is highest, so interaction with visible light is strongest. It is also the part that interacts strongest with the ground.

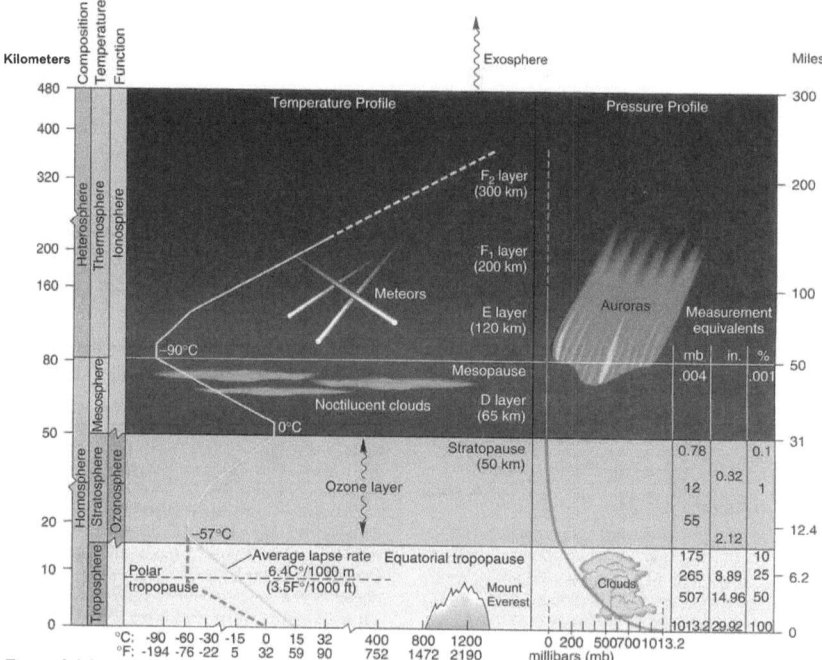

Figure 2.1.1
The different layers of the atmosphere and their specific characteristics. From Christopherson (1994).

Due to its temperature and pressure differences the atmosphere can be divided into several layers. This is demonstrated in Figure 2.1.1. However the intensive discussion of radiation permeation in the atmosphere is not part of the thesis. For a closer discussion of the topic I suggest as reference e.g. the book of Platt & Stutz (2008), Seinfeld & Pandis (2006) or *Light Scattering on small particles* from van de Hulst (1981). The following chapters just give some basic ideas of the influences on our measurements.

2.1.1 Absorption

Absorption is a process where incoming energy is conveyed from a photon to matter.

2.1.1.1 Fraunhofer lines

The light coming from the sun to the atmosphere can be described as approximately the radiation of a black body of 5800 K but in fact this light shows a lot of fine structures of extinction. They were first discovered by Joseph von Fraunhofer. These lines are caused by absorption processes in the sun. The Planck function for a black body of 5800 K and the solar spectrum is shown in Figure 2.1.1. The Fraunhofer lines are well visible in the green spectrum.

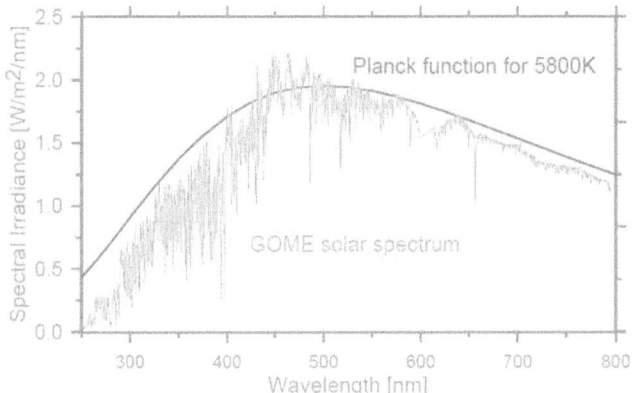

Figure 2.1.1. Comparison of a GOME solar spectrum (in green) to the Planck function for 5800K (in red). From Wagner et al. (2008 b).

2.1.1.2 Chemical Species in the atmosphere

As discussed before, the incoming light from the sun is not the perfect radiation of a black body. However this only explains the extinction until it reaches the top of the atmosphere. Within the atmosphere there are different chemical species which absorb solar light. These chemical species produce even more distinct features in the radiation spectrum reaching the ground. The strongest absorbers in the wavelength range studied in this thesis are O_2 and water vapour. Since these chemical species in the atmosphere have to be passed twice by the light, before it is detected by the satellite instrument, these absorption processes have to be very carefully considered. Some of the major species and their effect on the light spectrum are shown in Figure 2.1.2.

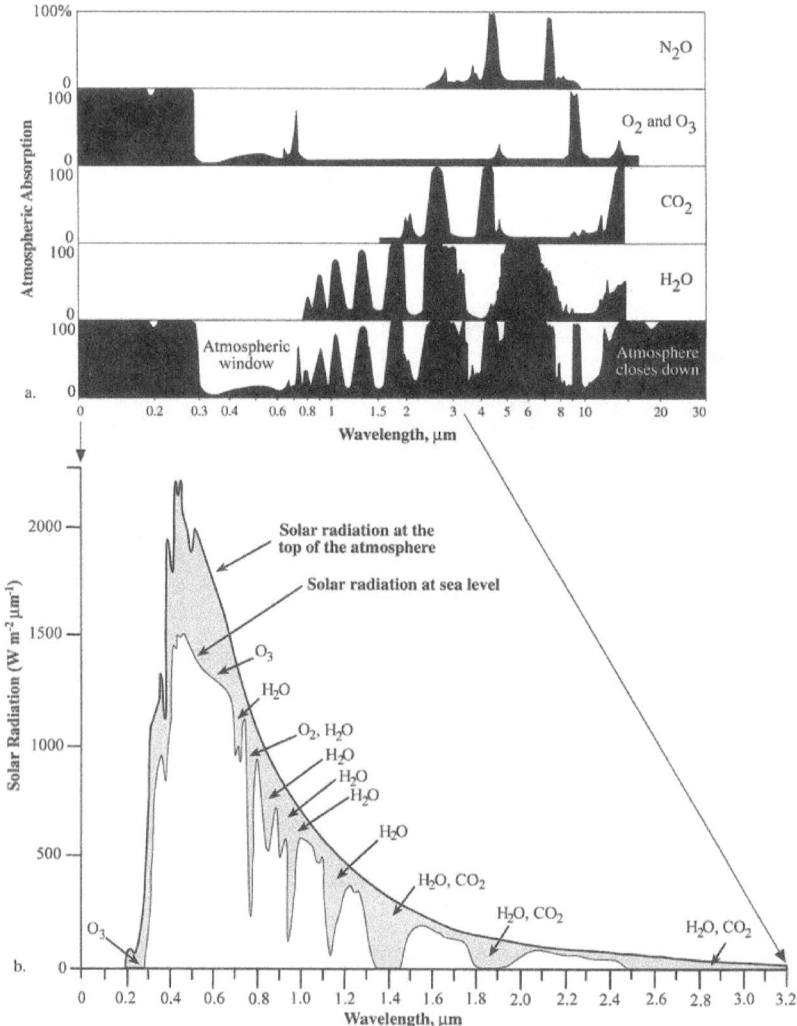

Figure 2.1.2
a) The absorption of the Sun's incident electromagnetic energy in the region from 0.1-30 μm by various atmospheric gases. The first four graphs depict the absorption characteristics of N_2O, O_2 and O_3, CO_2 and H_2O, while the lowest graphic depicts the cumulative result of having all these constituents in the atmosphere at one time. The atmosphere especially "closes down" in certain portions of the spectrum while "atmospheric windows" exist in other regions that transmit incident energy effectively to the ground. It is within these windows that remote sensing systems can be applied. b) The combined effects of atmospheric absorption, scattering and reflectance reduce the amount of solar irradiance reaching the Earth's surface at sea level (after Slatter, 1980) from Jensen 2000.

2.1.2 Scattering

There are basically two processes of scattering in the atmosphere: elastic and inelastic scattering. Elastic scattering is the name of the processes where incoming energy is redirected in its propagation but the energy stays the same. The wavelength is not changed. There are two different variations of elastic scattering : scattering at air molecules (Rayleigh-scattering) or scattering at aerosol particles (Mie-scattering). Inelastic scattering does have an effect on the wavelength. These processes are discussed in more detail in the following sections.

2.1.2.1 Mie

Mie-scattering happens when the particle in the atmosphere, which is interfering with the electromagnetic wave, is of the same size or even bigger than the wavelength. Such particles scatter predominantly in the forward direction and only very small amounts of the light are redirected to the sides and to the back. The forward scattering tendency becomes more and more pronounced the bigger the particle gets compared to the wavelength. The kinds of particles most likely to produce Mie-scattering in the atmosphere are e.g. aerosols, water droplets and ice crystals.

Water droplets and ice crystals form clouds. Clouds are a very common feature in the earth's atmosphere. They appear when water condensates around cloud condensation nuclei (CCN). The effect on incoming radiation is a very strong scattering for all wavelengths in the visible and near-infrared wavelength range. This is why clouds appear as bright white features on satellite images.

Depending on the thickness of the cloud the light path inside the cloud and between multiple cloud-layers (if present) may become extremely long through multiple scattering. This increases the effects of absorption inside the cloud, but at the same time shields information from underneath the cloud. The absorptions of atmospheric absorbers above the clouds may be increased.

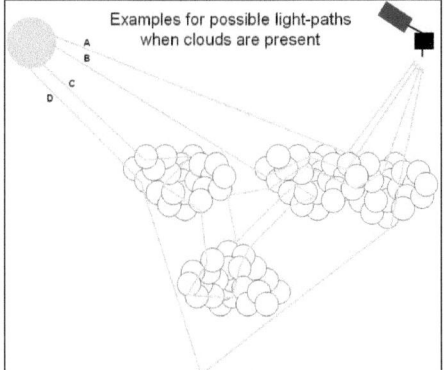

Since this thesis is analysing vegetation information, clouds will be mostly considered here as masking the desired signal. Therefore pixels of higher cloud-cover have to be removed. For further reference see Chapter 7.3.1.

Figure 2.1.6
Four examples for possible light-paths are given in the presence of clouds. A depicts scattering directly at the cloud-top, B shows multiple scattering inside the cloud. Scenario C shows the most complicated light-path, demonstrating multiple scattering inside clouds and between clouds. D shows light penetrating the cloud-layer, bearing information on scatterers on the ground.

Another constituent in the atmosphere, which strongly influences the light reaching the satellite, is the presence of aerosols. Aerosols may occur in very different layers of the atmosphere and may be caused naturally e.g. from volcanic eruptions or natural fires, but may also be caused by e.g. industry or manmade vegetation fires. Another natural source are dust storms originating in deserts. Aerosols may cause intensive scattering and absorption of electromagnetic radiation, depending on the chemical composition of the aerosols. Dark aerosols (like e.g. soot or smoke) tend to absorb intensively. Considering the extraction of information on vegetation on the ground, aerosols also are shielding the information from the satellite.

2.1.2.2 Rayleigh

Rayleigh-scattering occurs among the molecules in the atmosphere. The scattering object is small compared to the wavelength of the incident light. Rayleigh-scattering favours the forward and backward direction for scattering.

It is strongest at short wavelengths. The dependency between wavelength (λ) and strength of Rayleigh-scattering is $1/\lambda^4$. The wavelength dependence of Rayleigh-scattering become apparent, when looking at the sky on a clear sky day. The blue colour intensifies when looking away from the sun.

2.1.2.3 Raman scattering and Ring-effect

Raman-scattering is inelastic scattering, which means that the incoming energy is altered before it is sent off into another direction. The wavelength is either shortened or prolonged by a small margin. The resulting effect is that the radiation in the original wavelength is reduced and in the neighbouring wavelengths is increased.

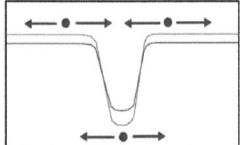

Figure 2.1.5
Filling in of Fraunhofer lines through inelastic scattering (Ring-effect).
Source: AT2-ELS

This becomes especially apparent in the very narrow extinction lines of the Fraunhofer lines. An example is given in Figure 2.1.5. Here the Raman-scattering into the Fraunhofer line is a lot more likely than out of it, since the incident number of photons inside the Fraunhofer line is reduced compared to photons at slightly different wavelengths. This causes a filling in of the Fraunhofer lines. This is called the Ring-effect because it was first discovered by Grainger and Ring (1962).

2.2 Vegetation

Light interacts with vegetation. It might be reflected, absorbed or transmitted. Depending on the wavelength range the dominant effect varies. This is demonstrated in Figure 2.2.1., where the upper curve shows reflectance, the lower curve shows transmittance and the gray areas in-between denotes absorptance.

In the longer wavelength range most of the absorption comes from water, cellulose and lignin. The different absorbers are shown in Figure 2.2.2. This wavelength-range is not

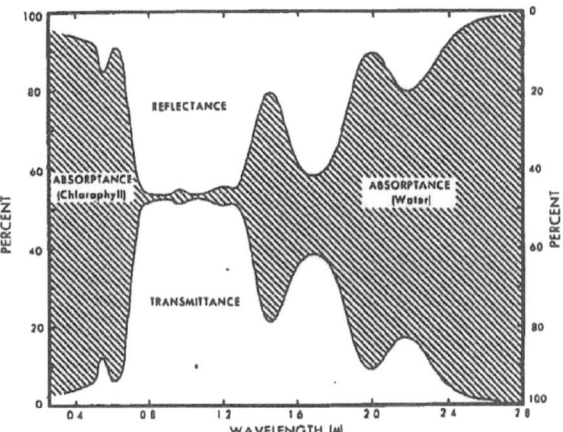

Figure 2.2.1:
Reflectance, absorptance and transmittance spectra of a plant leaf. (From Knipling 1970)

investigated here. More important for the currently interesting wavelength range of 500 to 800 nm is the absorption of pigments, which is for example demonstrated in Figure 2.2.3, where the contrast of reflectance of a white part of a leaf is shown against a green part. In the white part there is nearly no pigmentation whereas in the green part there is natural pigmentation.

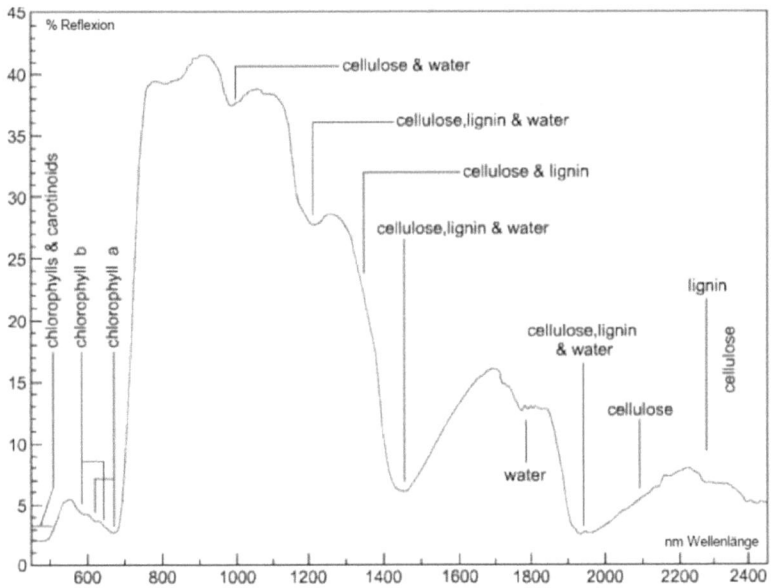

Figure 2.2.2
Major absorbers influencing the reflectance of vegetation. (From. M. Rast 1991, source JPL)

15

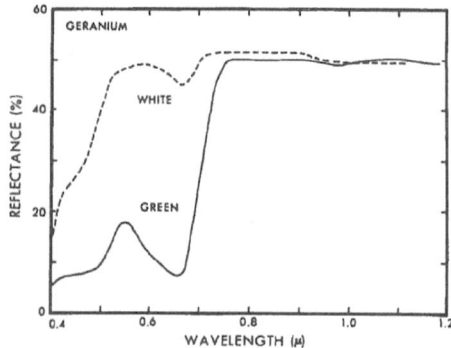

Figure 2.2.3.
Reflectance of the white and green portions of a vareagated geranium leaf. (From Knipling 1970)

The contrast between the white and the green leaf shows the enormous importance of pigments to vegetation-reflection in the visible light range. That is why this chapter first takes a closer look at the absorption of pigments. But even in the white leaf some structures are apparent in the spectral response. Those have to be due to other factors, like internal structure of the leaf or characteristics of plants and vegetation-stands as a whole. Those effects are explored later in this chapter.

2.2.1 Absorption by Pigments

"Plants exhibit two fundamentally different types of colour, which are due respectively to structural effects of the tissue and to the presence of pigments. These two forms of colouration may, and often do, co-exist in the same tissues, and the perceived colour will thus depend on the joint effect of two distinct phenomena. Many pigments have quite different colour characteristics depending on the environment in which they find themselves: in fat globules, aqueous suspension, pH, protein or other complexes." (Davies, 2004)

A chemical compound (including pigments) appears coloured when it selectively absorbs specific wavelength in the visible spectrum of light. This selective absorption in the chromophore group is essentially influenced by double bonds (like e.g. >C=C<, >C=O, >C=N-, or -N=N-); especially in conjugated systems (Richter, 1988; Wild et al., 1993). For such molecules, the energy required for excitation of the electrons to a higher energy level is reduced, allowing the molecule to be energised by light within the visible range.

All pigments have their own characteristic absorption maxima. Some of these are demonstrated in Fig. 2.2.4. It is apparent that the maxima are distributed all over the visible light range. The different maxima complement each other for optimal energy collection. The most important plant-pigments are chlorophylls, carotenoids and flavonoids, especially anthocyanins. The different groups of pigments all absorb in different wavelength-ranges, depending on their function. Since anthocyanins are often produced to protect tissue from damage by excessive energy-uptake, these pigments absorb mostly the high-energy short wavelengths of blue and UV light.

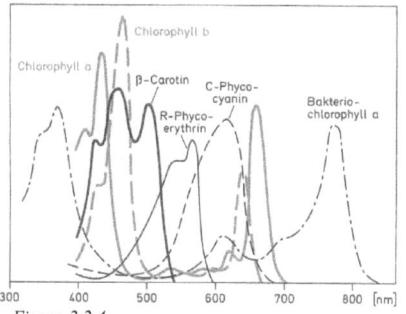

Figure 2.2.4
Absorption-spectra of some biologically important pigments. (Chlorophylls and β-Carotene in organic solution, others in watery solution.) From Strasburger (1991)

Chlorophylls are the most important pigments for photosynthesis. They convert the energy of blue and red light into an energy-form usable for plant physiology. They transport this energy by electron-donation. Carotenoids help capture light-energy of wavelengths not covered by chlorophylls. Depending on their position inside the tissue, they either transport this energy towards chlorophyll a for photosynthesis or fix it to protect against photo-oxidation.

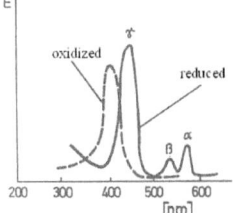

Unfortunately, the maximum absorption in natural pigments is not fixed at a certain position. Especially the pigments involved in photosynthesis exchange electrons. During this process they become reversibly chemically altered and consequently shift their maximum absorption. This is illustrated in fig. 2.2.5.

Figure 2.2.5.
Example for changing absorption due to electron-transport. Cytochroms are part of the photosynthetic electron transport and are localised in the chloroplast. They are built similar to chlorophylls except that the central Mg-atom is replaced by Fe. During electron-transport the central atom changes in electronic load (Fe^{3+}/Fe^{2+}). As a result the absorption changes between oxidized state (dashed line) and reduced state (solid line). From Strasburger (1991)

2.2.1.1 Chlorophyll

"The chlorophylls are omnipresent in photosynthetic tissue where they are responsible for photoreception (Formaggio et al., 2001). These tetrapyrrole pigments contain a centrally placed chelated magnesium ion and have typical green hues, although the presence of other pigments can mask their presence in intact tissue (Schoefs, 2002). The structures of chlorophyll a and chlorophyll b are shown in [Fig. 2.2.6]. These pigments differ only in that one methyl group of chlorophyll a is replaced by a formyl group in chlorophyll b" (Davies, 2004)

In vivo chlorophylls are surrounded by different chemical environments depending on the kind of plant, algae or bacteria in which they are active. Different proteins and amino-acids near the photosynthetic units exert different attractions to the ends of the long chlorophyll molecule. This may cause partial charge and bending of the molecule. As result a shift in maximum absorption may occur. Examples of this are given in Fig. 2.2.7.

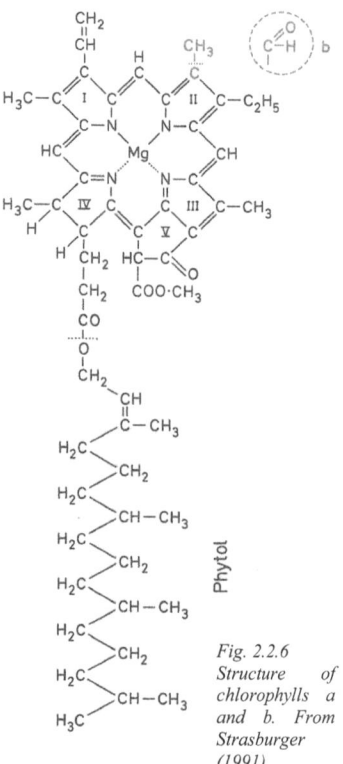

Fig. 2.2.6 Structure of chlorophylls a and b. From Strasburger (1991)

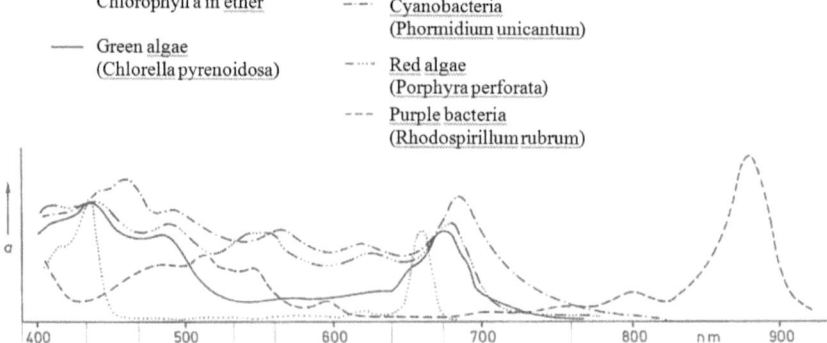

Fig. 2.2.7
Comparison of absorption-spectra of a solution of chlorophyll a and different photosynthetic Organisms. From Richter (1988)

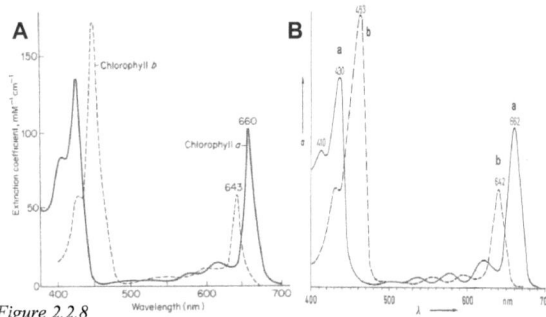

Figure 2.2.8
Absorption spectra of chlorophylls extracted in ether. Note the differences between the results from A) Hall and Rao (1999) and B) Richter, G. (1988)

Even within the same environment the absorption peaks of chlorophyll may alter as demonstrated by Fig. 2.2.8., where the measurements of Hall and Rao and Richter are compared. Although chlorophylls a and b were extracted in ether in both cases, the maximum absorption differs by up to two nanometers.

Strasburger (1991) mentions chlorophyll-a-molecules that undergo reversible changes in absorption when they donate an electron for a short time during light-harvesting, thus playing a central role during photosynthetic electron-transport.

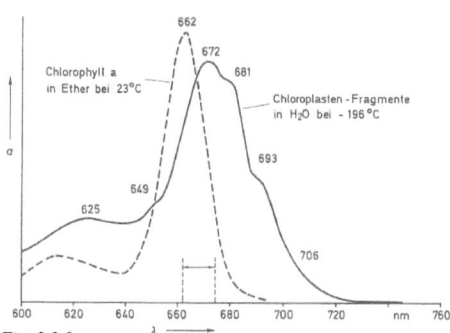

Fig. 2.2.9
Different absorption of red light by chlorophyll a in solution (solvent: ether) and connected to the structure of a chloroplast (measured from a fragment of Euglena). In the last case the maximum experiences a red-shift of 10 nm and also displays at least five different sub-maxima as shoulders to the peak; to improve registration, the last measurement was done at very low temperatures. Translated from Richter (1988)

But not only the chemical environment influences the absorption. Chlorophylls are extremely important for plant nutrition and to enable their function they are bound into complex structures. They are bound into light-harvesting complexes (LHC) and photosystems I and II. In all those structures chlorophylls and other pigments are bound in very close proximity, so that dipole-dipole interactions and exchange interactions become possible. Those processes lower the energy-level and cause a red-shift of the absorption-maxima. This is extremely important to allow efficient photosynthesis. In addition these complexes are collected on the Thylakoid membrane in the chloroplasts. This membrane is then folded into grana, bringing all these complexes very close together, allowing even more interactions. This has an influence on the absorption in the living tissue. This is demonstrated in Fig. 2.2.9.

All these effects contribute to the phenomenon that chlorophyll absorption maxima vary during measurements in the field.

Other variabilities in chlorophyll-signal may be caused by variable ratios of chlorophylls a and b or by physiological alterations: "The chlorophyll a/b ratio varies from 2.0–2.8 for shade-adapted plants to 3.5–4.9 for plants adapted to full-sun conditions. This variation in chlorophyll a/b ratios is due to differences in the ratio of photosystem I (PSI) to photosystem II (PSII) and the size and composition of the light-harvesting complexes (LHCs) associated with each photosystem. The photosystems contain chlorophyll a but not chlorophyll b, whereas the LHCs contain significant amounts of chlorophyll b."(Davies, 2004)

Shade-adapted plants have lower chlorophyll a/b ratios than sun-adapted plants because they tend to have more LHCs associated with their photosystems than sun-adapted plants (Anderson, 1986; Porra, 2002).

In vivo a number of spectral variants of chlorophyll a are also detected. These spectral differences result from the different environments within LHCs and photosystems. Other minor chlorophylls and chlorophyll derivatives have also been reported, with the most important ones being chlorophyll a0, found in the PSI reaction centre, and pheophytin, found in the PSII reaction centre. Other forms of chlorophyll except a and b may be found in algae and bacteria (Willows, 2003). Those are not relevant for this thesis.

The photochemical properties that allow chlorophylls to carry out their function in photosynthesis also present potential problems for plants. Excessive absorbed light energy that is not used in photosynthesis, must be dissipated in some way. This dissipation can occur by a number of mechanisms, including fluorescence and reaction with other compounds. The potential of excited-state chlorophylls to react with oxygen to form singlet oxygen is one of the most damaging for the plant (Reinbothe et al., 1996; Matile et al., 1999; von Wettstein, 2000). The generation of singlet oxygen can possibly lead to cell death by a cascade of free radical mediated reactions forming a variety of reactive oxygen species that can damage proteins and nucleic acids. Plants have developed mechanisms to limit the formation of singlet oxygen and free radicals to prevent these damaging effects.

"Some of these protective mechanisms include the use of antioxidants such as ascorbic acid and a-tocopherol; expression of superoxide dismutase and peroxidases; a reduction or increase in the total amount of chlorophyll by modifying the size of the chlorophyll antennae when grown under high or low light conditions, respectively; modification and use of accessory pigments, such as carotenoids via the xanthophyll cycle, to dissipate excess light energy from chlorophylls before reaction with oxygen can occur." (Davies, 2004)

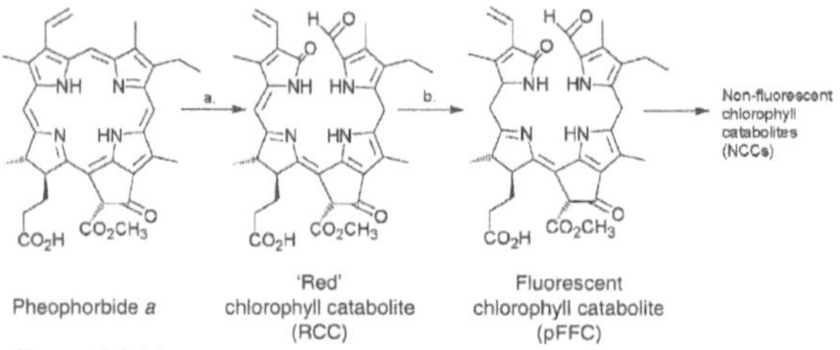

When chlorophyll is metabolised, the resulting chemical structures create distinct absorption-characteristics of their own, as shown by Figure 2.2.10. These coloured biosynthetic intermediates are tightly controlled because an excessive accumulation may cause phototoxic effects.

Figure 2.2.10
Key steps in chlorophyll degradation: (a) pheophorbide a oxygenase; (b) RCC reductase. From Davies (2004)

Pheophorbide a → 'Red' chlorophyll catabolite (RCC) → Fluorescent chlorophyll catabolite (pFFC) → Non-fluorescent chlorophyll catabolites (NCCs)

"The half-life of chlorophyll within a normal plant leaf is estimated to be between 6 and 50 h. In addition, during the highly visible colour change which occurs in autumn in deciduous plants and also in other senescing processes, chlorophyll is also degraded. The main reason for chlorophyll degradation during senescence is so that the plant can recover nutrients, such as the nitrogen tied up in the proteins of the photosynthetic proteins, without having to risk potential photo-oxidation caused by the chlorophyll that would be liberated by this process." (Davies, 2004, p41)

All these described phenomena alter the chlorophyll reflection-signal which is obtained from spectral measurements over vegetation.

2.2.1.2 Carotenoids

Carotenoids are crucial pigments for photosynthesis. They must be present in all photosynthetic tissue because they are involved in photosystem assembly, responsible for protecting the chlorophylls from photodynamic destruction and in addition function as auxiliary pigments (Young & Frank, 1995). They do that by contributing to light harvesting by absorbing light energy in a region of the visible spectrum where chlorophyll absorption is lower and by transferring the energy to chlorophyll.

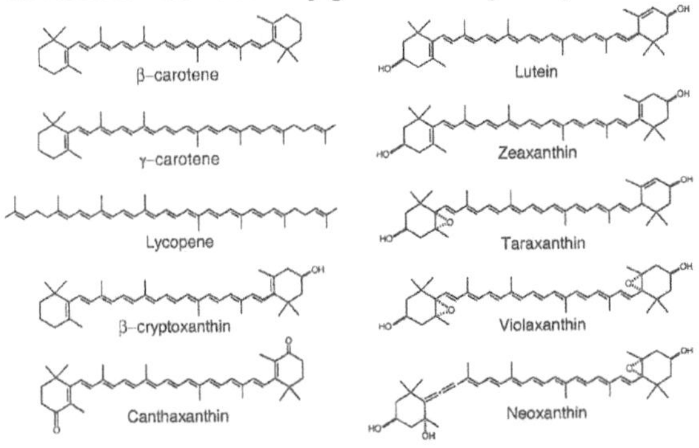

Fig. 2.2.11.
Structures of some typical carotenoids. From Davies (2004)

The synthesis of carotenoids and chlorophylls and their subsequent binding to pigment-binding proteins must be precisely balanced to avoid extensive photo-oxidative damage and to meet the appropriate photosynthetic demands of the various growth conditions that plants are exposed to on a daily and seasonal basis (Herrin et al., 1992; Anderson et al., 1995). Thus, carotenoid transcriptional regulation and therefore chloroplast–nuclear signalling must be responsive to environmental stimuli, oxidative stress, redox poise and metabolite feedback regulation. The nature and mechanisms of such signals is still under investigation.

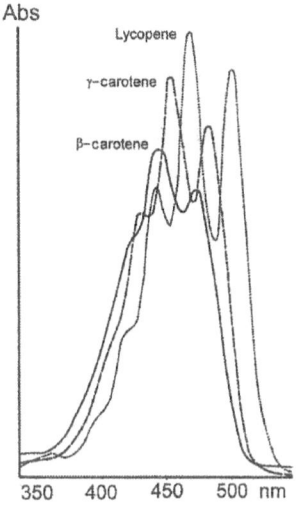

Fig. 2.2.12
Visible absorption spectra of β-carotene, γ-carotene and lycopene in petroleum ether. Adapted from Vetter et al., (1971)

"Carotenoids, named originally because of their isolation from carrots (Dauca carota), are yellow to red-coloured terpenoids containing eight isoprene units arranged in a symmetrical linear pattern. The chain itself contains a series of conjugated double bonds that are responsible for the typical visible light absorption of this class of compounds. The number of double bonds is normally nine or more, although shorter chromophores are occasionally found. There is little skeletal variation, and it is almost entirely restricted to cyclisation at the end(s) of the chain where simple functio-nalisation can also occur." (Davies, 2004) Figure 2.2.11 illustrates some typical structures of carotenoids. It can be seen that only oxygen functions are found, e.g. as in ketones, alcohols, epoxides. These oxygenated compounds are sometimes known as xanthophylls to distinguish them from the hydrocarbon carotenes.

21

Figure 2.2.12 shows some absorption spectra of carotenes. All display several absorption maxima within about 100 nm, some peaks absorbing at 500 nm and above, which makes them interesting for this thesis. Most carotenes have distinct absorption characteristics. This is why much of the early work on the identification of carotenoids was based on their characteristic visible light absorption spectra, and this technique is still widely applied (Britton, 1995).

"The two major classes of carotenoids are the carotenes and their oxygenated derivatives, the xanthophylls. The most abundant xanthophylls, lutein and violaxanthin, are key components of the light-harvesting complex (LHC) of leaves, and are responsible for the yellow colour of autumn leaves that is normally masked by the green chlorophylls (xanthos ¼ yellow and phyll ¼ leaf)." (Davies, 2004)

2.2.1.3 Flavonoids, especially Anthocyanins

Flavonoids serve many different functions in plants. They are used for signalling to microorganisms, protection against pathogens, amelioration of biotic and abiotic stress, influencing auxin transport and enabling plant fertility. But their most obvious and best characterised role is to provide floral visual cues for insect and animal pollinators. This makes them the most important floral pigments, occurring throughout the angiosperms and providing most colours in the visible spectrum. Flower colour and therefore flavonoids are among the oldest subjects in formal plant science. In 1664, it was found that the purple pigment from Viola changed colour between red and blue depending on the acidity of the solution and therefore was a natural pH indicator (Boyle, 1664) and in 1835 the term 'anthocyan' was coined for the most important flavonoid pigment (Swain, 1976), a derivative of which is used today.

Flavonoids have a 15-carbon (C15) base structure comprised of two phenyl rings (called the A- and B-rings) connected by a three-carbon bridge that usually forms a third ring (called the C-ring). The degree of oxidation of the C-ring determines the various classes of flavonoids. Although they are very similar in structure, only some flavonoids have the ability to absorb visible light and are thus pigments. The two key structural features of a flavonoid pigment are the degrees of double bond conjugation and oxygenation (in the form of hydroxylation). Increasing either causes light at longer wavelengths to be absorbed. The predominant flavonoid pigments are the anthocyanins, while chalcones, aurones and flavonols serve a more limited role (Davies, 2004,).

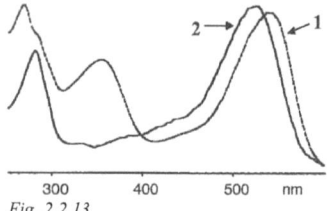

Fig. 2.2.13
The UV–Vis absorption spectra of the covalent anthocyanin–flavonol complex, (cyanidin 3-O-bglucoside)(kaempferol 3-O-(2-O-b-glucosyl-b-glucoside)-7-O-b-glucosiduronic acid) malonate (1), and theanthocyanin cyanidin 3-O-b-glucoside (2) (Fossen et al., 2000). The bathochromic shift (16 nm) of the visible absorption maximum of 1 compared to 2 indicates intramolecular association between the anthocyanin and flavonol units of 1. From Davies (2004)

In contrast to other flavonoids, anthocyanins absorb light in the visible area around 520 nm under acidic conditions (Fig. 2.2.13). For characterisation purposes under controlled solvent conditions, the absorption maximum in this region is related to the nature of the aglycone, position of sugar substituents, presence of aromatic acyl groups, co-pigmentation, self-association, etc. (Strack and Wray, 1989).

The ability of anthocyanins to absorb the UV is especially important for young, developing plant-tissues. Böhlmann (2009) points out that many plants protect their young shoots and developing leaves with anthocyanins until the cells are mature enough to use the incoming light for photosynthesis. When the high-energy light is no longer dangerous for the organism, the protective pigments are broken down.

In many species anthocyanins readily accumulate on exposure to UV-B. These compounds are coloured and absorb up to 530 nm. Consequently, they do not provide protection per se (Solovchenko & Schmitz-Eiberger, 2003), but may be esterified to cinnamic acid to modify their absorption spectrum (Jordan, 1996).

Inside the cell, anthocyanins are deposited in the vacuole. In some species, highly pigmented bodies, which have been termed anthocyanoplasts, may form in the vacuole (Pecket & Small, 1980) or AVIs (Markham et al., 2000).

Anthocyanins absorb between 270 and 290 nm in the UV and 465 and 560 nm in the visible light. The second absorption-range reaches into the wavelength range investigated in this thesis. In both cases the high seasonal and situational variablilty in concentration should be considered.

2.2.2 Scattering inside the leaf

„The characteristics of light reflectance and transmittance can be explained mainly on the basis of critical reflection of visible light at the cell wall-air interface of both the palisade and spongy layers of the mesophyll [Figure 2.2.14]. It has been shown that reflectance increases with an increase in the number of intercellular air spaces. This is because diffused light passes more often from highly refractive hydrated cell walls to lowly refractive intercellular air spaces." (Barrett & Curtis, 1992)

The leaf structure is important in that the upper and lower epidermal layers together with the palisade cells and spongy mesophyll each play a part. Palisade cells are usually directed towards the incoming light. They contain the highest concentration of absorbing pigments. Spongy mesophyll is important for the plant since it is needed for the sufficient exchange of gasses for physiology and is important for radiative transfer because it scatters near-infrared light. Barrett and Curtis (1992) explain that often the leaf becomes more spongy with age with the result that the mature leaf displays less reflectance in the visible bands (about -5%) and more in the infrared (about +15%).

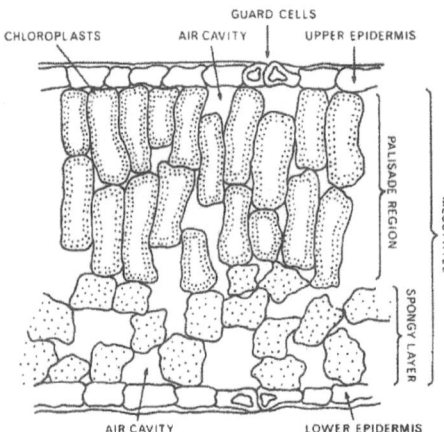

Figure 2.2.14:
A generalized diagram of leaf structure. (From: Curtis, 1978)
In the palisade region most of the photosynthetically active organelles are located. The major function of the spongy layer is gas-transport, providing CO_2 for photosynthesis and venting O_2 back to the atmosphere.

"Frequently the different parts of the spectrum are affected differently by variations in plant composition and structure. The 0.5-0.75 µm band is characterized by absorption by pigments consisting mainly of chlorophylls a and b, carotenes and xantophylls. The 0.75-1.35 µm band is a region of high reflectance and low absorption which is greatly affected by the internal leaf structure. The 1.35-2.5 µm band is influenced somewhat by internal structure but is more particularly affected by water concentration in the tissue. In general the spectral transmittance curves for mature and healthy leaves are similar to their spectral reflectance curves for the 0.5-2.5 µm bands but are slightly lower in magnitude." (Barrett & Curtis, 1992)

2.2.2.1 Build-up of a cell and the influence on scattering properties

Differences of refractive index occur not only between cell walls and intercellular air spaces, but also within the cell. "Refractive index discontinuities among membranes, cell walls and protoplasts induced near-infrared light scattering." (Gausman, 1977) The scattering within the cell is estimated to contribute about 8% of the reflection at a wavelength of 800 nm. This is only an estimate, because depending on the age of the cell, the internal structures are greatly different. This can be seen in Figure 2.2.15, which shows that an embryonic cell is well filled with cytoplasm and organelles while a mature cell often pushes the cytoplasm to the walls and is dominated by the huge internal vacuole. The vacuole is filled with a watery solution of enzymes and other molecules, depending on species and situation of the plant.

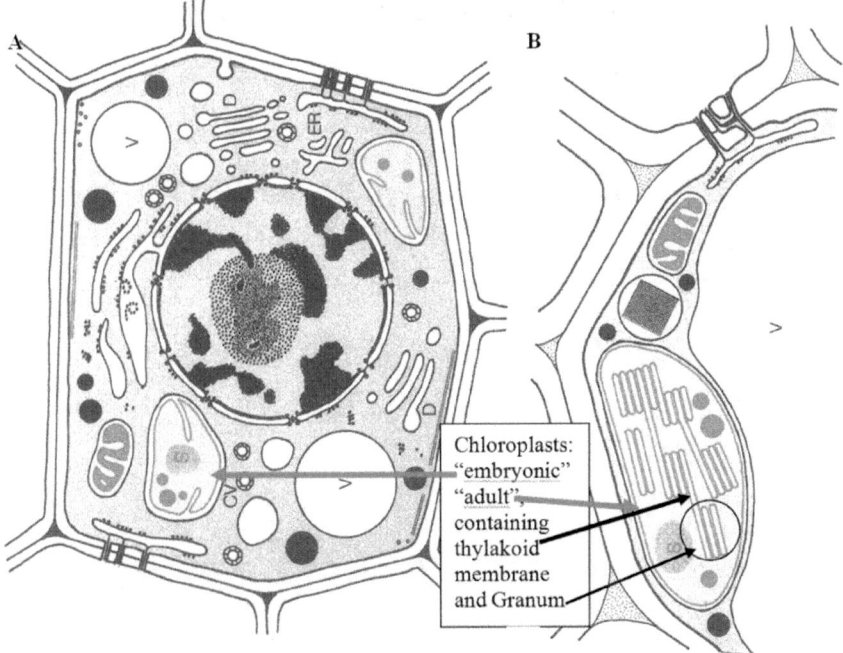

*Figure 2.2.15: Build-up of plant-cells of different ages: **A**, embryonic cell, **B**, adult cell e.g. in a leaf. V: Vacuole, S: accumulation of starch, CV: coated vesicles, ER: endoplasmatic Reticulum, solid circles: oil-droplets (refraction!) (adapted from: Strasburger 1991, p22)*

2.2.2.2 Build-up of a leaf and the influence on scattering properties

Leaves are a central provider of plant survival and are therefore highly organized organisms, which mostly results in a layered structure of an outer epidermis, palisade layer, spongy mesophyll layer and a lower epidermis. Most of the photosynthesis takes part in the palisade layer where there is the highest content of chloroplasts, chlorophyll and other pigments. The basic elements of the structured leaf are shown as a three-dimensional block-diagram in Figure 2.2.16.

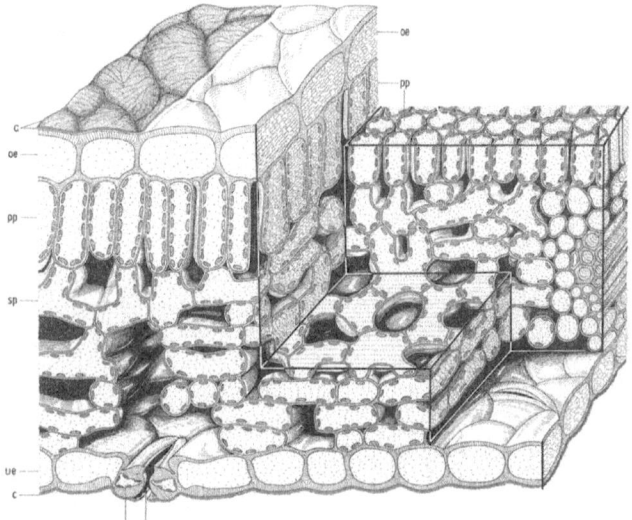

Figure 2.2.16:
The build-up of a deciduous leaf: c: cuticule (red), oe: upper epidermis, pp: palisade layer, sp: spongy layer, sz: guard cell, ue: lower epidermis (From Nultsch 1986)

Most of these different structures have different influences on reflection, as is stated by Knipling (1970): "The cuticular wax on a leaf is nearly transparent to visible and infrared radiation, and very little of the solar energy incident to a leaf is reflected directly from its outer surface. The radiation is diffused and scattered through the cuticle and epidermis to the mesophyll cells and air cavities in the interior of the leaf. Here the radiation is further scattered as it undergoes multiple reflections and refractions where refractive index differences between air (1.0) and hydrated cellulose walls (1.4) occur."

During the senescence of a leaf structures may change, especially when a leaf gets old or if disease strikes. The water content within the different cells may shrink, reducing the pressure inside the cell, allowing cell-membranes to fold and thus collapse air spaces. Having fewer airspaces reduces the amount of differences in refraction, therefore bringing down the infrared radiation. The effect of onset of disease on reflection is demonstrated by Figure 2.2.17.

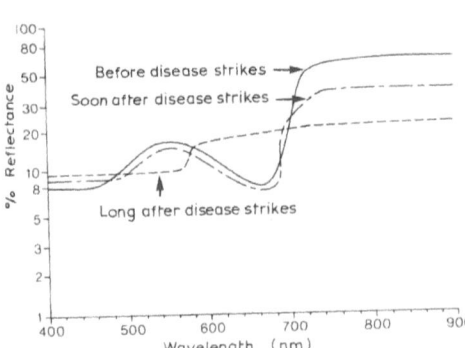

Figure 2.2.17:
Reflectance characteristics of healthy and diseased leaves. (From Barrett & Curtis 1992)

Another interfering factor may be strong solar radiation, especially in the UV-B range. Too much UV-B may cause multiple reactions in leaves, from modified cellular organisation and production of additional pigments (Flavonoids) to changes in cell metabolism. Figure 2.2.18 illustrates the different protective mechanisms against UV-B.

Protective mechanisms against UV-B

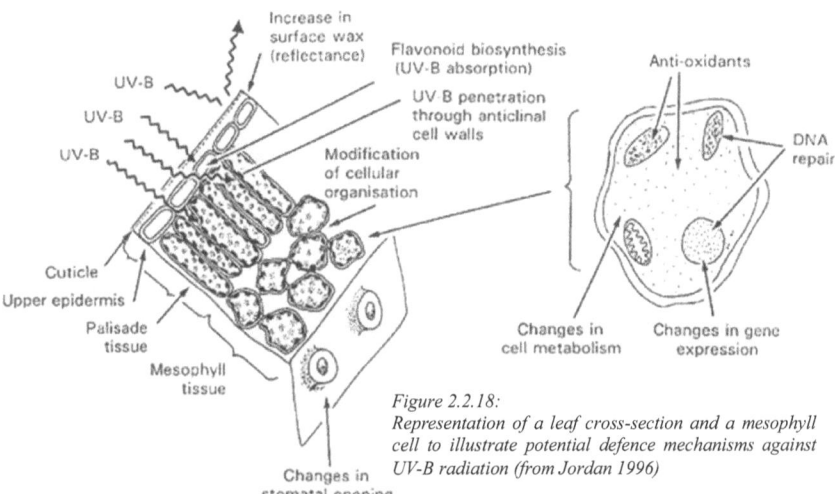

Figure 2.2.18:
Representation of a leaf cross-section and a mesophyll cell to illustrate potential defence mechanisms against UV-B radiation (from Jordan 1996)

Lack of water may be fended off by xeromorph leaves, either through a special built in the leaf cross-section or by rolling the leaf into protective position when necessary. These mechanisms are illustrated in Figure 2.2.19. The special leaf cross-section in **A** is an increased number of layers of epidermis, providing better protection for photosynthetically active layers underneath and protective structures for guarding cells, reducing dangerous loss of water-vapour. Evaporation is intensified by exposure to wind/air movement. This is countered by placing a guard cell inside a cavity. Hair inside the cavity enhance the protection. The mechanism in B and C exploits the same effects, producing the protective cavity by active movement of the leaf. This allows higher flexibility to a changing environment.

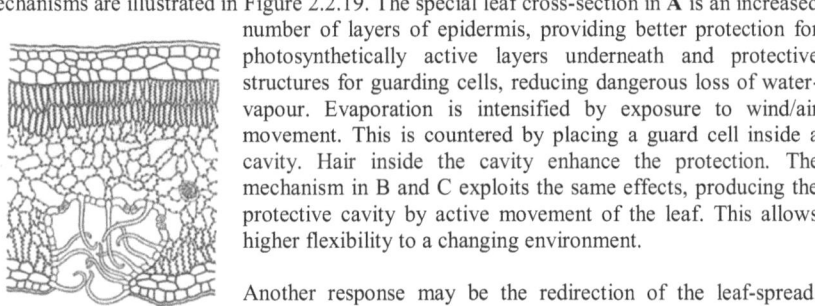

Another response may be the redirection of the leaf-spread. While in cool conditions the leaves are turned for maximum exposure to light, in hot and dry conditions the same tree may orient its leaves vertical to the light, thus minimising incident radiation. All these mechanisms influence the light-reflection-signal from vegetation picked up by satellites.

Figure 2.2.19:
Cross-sections of xeromorph leaves. *A*, oleander, e: multi-layered epidermis, p: two layers of palisades, t: thick spongy layer, guard cells are deep inside a cavity where air convection is severely reduced by hair. **B**, **C**, Stipa capillata, guard cells are limited to upper epidermis. In dry conditions the leaf rolls up, effectively shutting off access of guard cells to the atmosphere. When water is plenty, the leaf is spread flat.

2.2.3 Scattering within the canopy

Leaves don't only scatter and reflect light; they also transmit it, which has the result that since vegetation usually is a multi-layered environment, the amount of light penetrating to lower parts of the vegetation is reduced but not completely extinct. Thus the lower layers of the vegetation contribute to the total reflected signal.

The complexity of these light-interactions is hinted at in Figure 2.2.20. Here, the different effects of reflection and transmission on two layers of leaves are demonstrated, while ignoring absorption. Even this simplified model shows multiple interactions of the incoming light with the different leaf-layers. In a natural canopy these effects are multiplied.

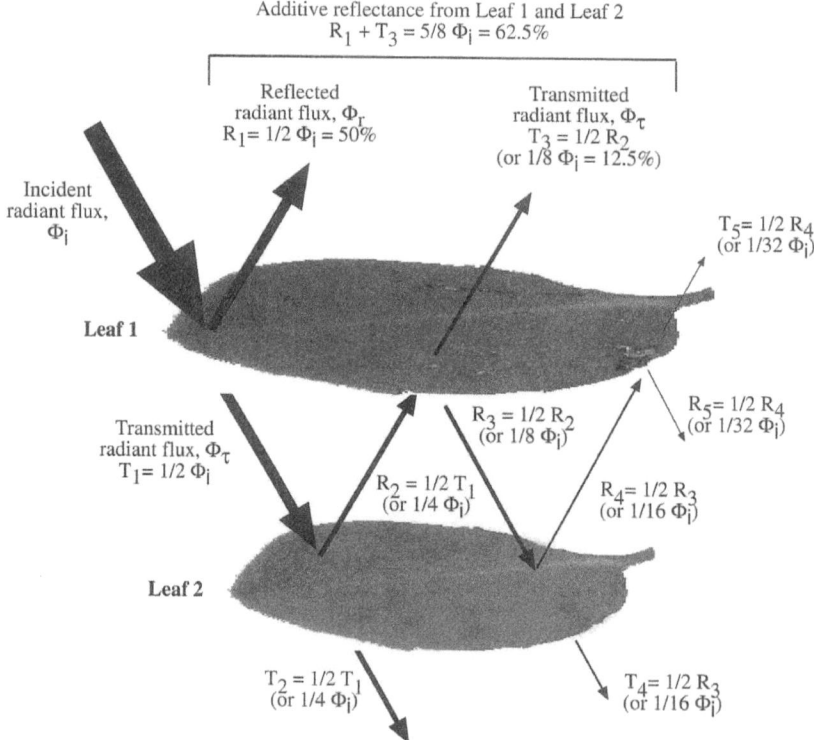

Figure 2.2.20
A hypothetical example of additive reflectance from a canopy with two leaf layers. Fifty percent of the incident radiation flux, Φ_i, to leaf 1 is reflected (R_1) and the other 50 percent is transmitted onto leaf 2 (T_1). Fifty percent of the radiant flux incident to leaf 2 is transmitted through leaf 2 (T_2), the other 50 percent is reflected toward the base of leaf 1(R_2). Fifty percent of the energy incident at the base of leaf 1 is transmitted through it (T_3) while the remaining 50 percent (R_3) is reflected toward leaf 2 once again. At this point, an additional 12.5 percent (1/8) reflectance has been contributed by leaf 2, bringing the total reflected radiant flux to 62.5 percent. However, to be even more accurate, one would have to also take into account the amount of energy reflected from the base of leaf 1 (R_3) onto leaf 2, the amount reflected from leaf 2 (R_4), and eventually transmitted through leaf 1 once again (T_5). This process would continue. From Jensen (2000)

There is a pronounced difference between forest areas and for example grassland, as demonstrated by Figure 2.2.21, showing the differences of the penetration depth and percentage of light contributing to total reflection.

The most apparent difference is the direct reflection, which in a forest area might be up to 10% and over grass about double that. Whereas in a forest stand penetration depth of light to the ground is expected to be of about 2%, in grassland the amount of light actually reaching the ground is about 5%. In forest areas the highest percentage of reflection of light penetration is into the upper canopy of the trees, where it is about 79%,. In contrast penetration of light relative to the vegetation stand is a lot deeper in grasslands, where the upper part only retains about 5%, while 36% still reaches down to about mid-level and another 34% reaches the layers just above the ground. So photosynthetic capabilities right down to the bottom of the plant make a lot more sense in grasslands than in forest areas. This is also demonstrated by plant habit. Trees growing inside stands do not contain branches and leaves in the lower areas. This may be very different for trees standing in the open.

Figure 2.2.21: Light-distribution in a layered mixed forest and in a grassland (altered from Larcher 1984)

Studying forest, it also has to be considered, that the light may come from different storeys of the forest depending on the season especially when it comes to deciduous forests where the leaves are changing with the seasons. Deciduous forests are completely without leaves during winter and early spring, when the light might penetrate all the way to the ground. This has the result that the vegetation signal is most probabaly due to understorey vegetation during early spring. In contrast when the leaves develop in later spring, summer and autumn, the signal should come mostly from treetops. In late autumn and early winter, before the snow comes, the vegetation signal may again be dominated by undergrowth. These changes in light penetration to the ground are demonstrated in Figure 2.2.22.

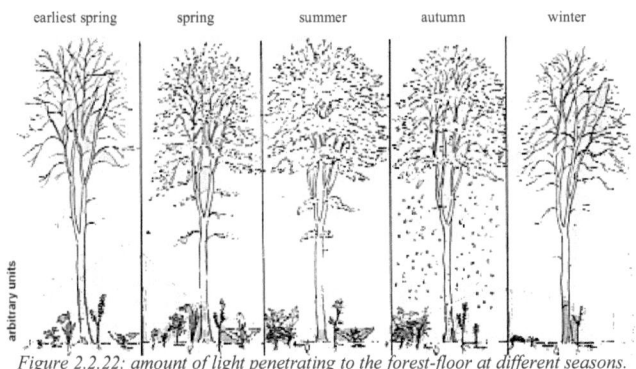

Figure 2.2.22: amount of light penetrating to the forest-floor at different seasons. Adapted from Hofmeister (1987)

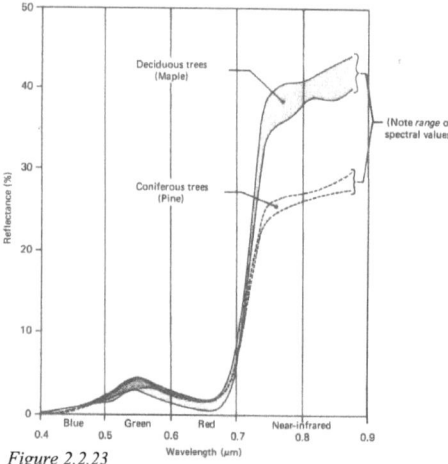

Another forest-phenomenon is the difference in reflectance of coniferous and deciduous trees in the near infrared (NIR) as shown in Figure 2.2.23. An explanation for this effect is suggested by Rautiainen (2005, p42): "the low NIR reflectances observed in coniferous areas are mainly due to within-shoot scattering." This takes into account that needles usually do not stand alone but grow in shoots (like in the two photographs of different pines), allowing multiple scattering between needles.

Figure 2.2.23
Generalized spectral reflectance envelopes for deciduous (broad-leaved) and coniferous (needle-bearing) trees. (Each tree type has a range of spectral reflectance values at any wavelength.) From Lillesand & Kiefer (1990)

Picture 2.1
Shoots of Pinus koraiensis (A) and Pinus peuce (B)

2.2.3 Differences in leaf-build-up and canopy

Leaf-build-up differs considerably, not only between different species of vegetation, but also within a single plant. This is for example demonstrated in the picture Figure 2.2.24. which demonstrates the different cross sections of leaves of the European beech depending on their exposure to light.

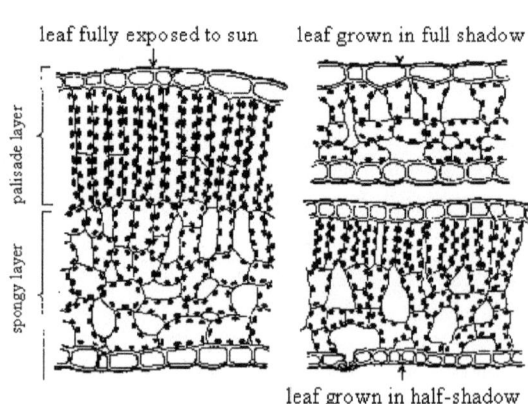

Figure 2.2.24:
cross-sections of leaves from the same Fagus sylvatica with different expositions to sunlight. (From Nachtigall 1985)

The leaf that is most strongly exposed to light has a very thick palisade layer, which contains the most photosynthetically active structures and also rather massive spongy mesophyll to supply sufficient amounts of gas exchange for the very intensive photosynthesis. In contrast, the leaf mostly developed in the full shadow of the canopy, is a lot thinner and contains only spongy mesophyll with very few amounts of photosynthetically active structures. Depending on the exposure to sunlight there are also intermediate stages.

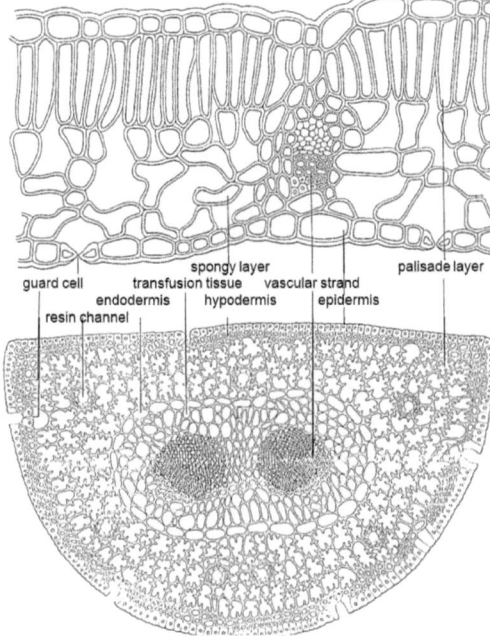

Figure 2.2.25:
Comparison of a bifacial deciduous leaf from Fagus sylvatica and a needle leaf of Pinus nigra. (From Böhlmann 2009)

These differences in leaf-build-up don't only occur within species but also between species, as for example demonstrated by Figure 2.2.25. Here the cross-section of a deciduous leaf is compared to the cross-section of a common needle leaf. The differences in build-up are quite pronounced. especially also showing that the number of intercellular air spaces may be a lot fewer in coniferous leaves than in deciduous leaves.

The general build-up may differ considerably. Schematic differences in leaf-build-up are shown in Figure 2.2.26. Here the general differences in leaf-build-up are compared between flat leaves, rounded leaves and intermediate leaf-build-ups. The grey area always presents the photosynthetically most active area of the leaf. Depending on the type of vegetation these most active areas are within the top or the bottom of the leaf or sometimes all the way around. However, these generalisations leave room for lots of differences within the single plants.

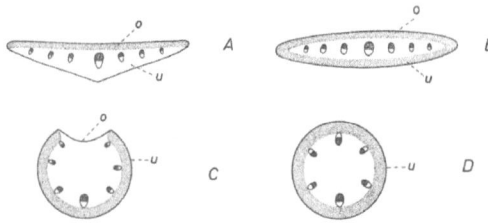

Figure 2.2.26:
Cross-sections of different leaf-types. A, dorsiventral, B equifazial, C, intermediate, D, uni-fazial. Assimilation layer dotted. o: upper epidermis, u: lower epidermis. (From Nultsch 1986)

In Figure 2.2.27 microscopic cross- sections for different needles of trees of the Pinaceae family are shown: five different cross- sections present five very different profiles and build-ups of the needles. Note that the leaf of Tsuga canadensis (E) has the most flattened-out structure and contains a big central area of air space whereas the needle of Pinus nigra (A) appears the most compact.

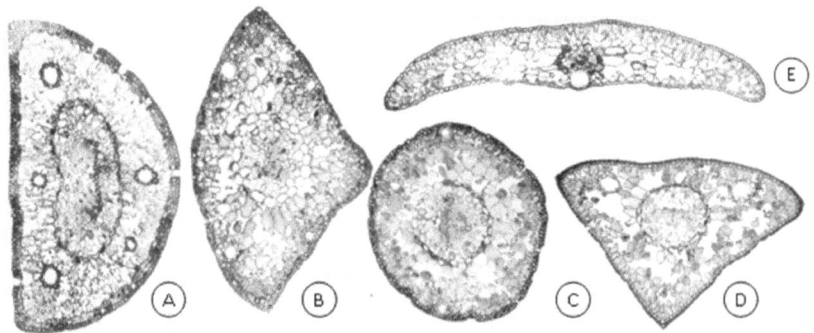

Figure 2.2.27
Cross-sections through various needles from Pinaceae-trees: A) Pinus nigra, B) Pinus bungeana, C) Pinus monophylla, D) Pinus cembra, E) Tsuga canadensis. Source: Böhlmann (2009)

In addition, there are also some differences in general cross- section build-up for broadleaf and grass vegetation.

Depending on the climate in which the vegetation grows, the process of photosynthesis might change. This basically happens in very hot and dry areas. Here parts of the process of photosynthesis, which are reliant on huge amounts of gas exchange, might be delayed into the night (CAM-cycle) or shifted into functional areas (C4-cycle), so that the plant doesn't lose too much water vapour during the gas exchange. This requires a different leaf-build-up, which is demonstrated by Figure 2.2.28. That might cause differences in reflectance in vegetation as well.

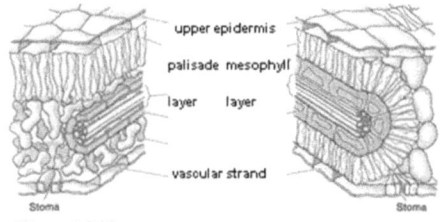

Figure 2.2.28:
Anatomy of a C3- and a C4-leaf. (Ffrom Böhlmann 2009)

Other adjustments especially to dry and hot environments are demonstrated in Figure 2.2.19 in Chapter 2.2.2.2, which shows that the guard cells for gas exchange might be situated within a cavity on the underside of the leaf, often even protected with additional hairs, producing different reflectance structures again in the leaf. Moreover even more important, some of the leaves may actively change their form, locking these cavities when it becomes too hot and dry outside to stop the gas exchange and to contain water vapour. This way the upper side, which usually is exposed to the sunlight, becomes rolled in and the underside of the leaf becomes the most visible surface, e.g.for reflection purposes and for reception of satellite information.

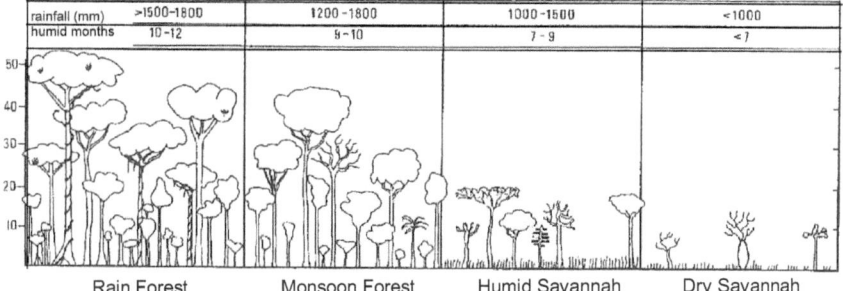

Figure 2.2.29
From Rain Forest to Dry Savannah vegetation becomes successively more sparse. Trees grow less high and start throwing off leaves with the season (shown as trees with apparent branches). Gras only grows in the understorey of Savannahs. Reduced rainfall and reduced number of humid months also reduce height and density of grasses. From Bremer (1999)

So far, differences shown have been on the scale of individual leaves. But changes in canopy and vegetation occur often according to local situations, sometimes on a very small scale, depending on the structure of the soil, elevation of the land, exposure to different gradients of the ground and also according to reception or availability of water. These changes are demonstrated for different forest classes in Figure 2.2.29, showing the different possibilities in tropical areas, depending on rainfall. These changes might occur over very small areas - often within a single pixel of the satellites we are using.

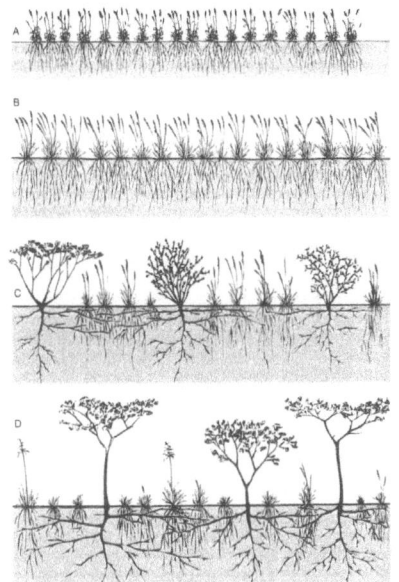

This also is important for grasslands. Due to the availability of water and nutrients, grasses might produce a very thin cover of only about 20-30 cm, or if the conditions improve, grass may grow up to two or more metres high. When the conditions are favourable, then the grassland may be interrupted by bushes and trees.

The naming of grasslands with scattered-in bushes and trees is not clearly defined. One such naming-system is presented in Figure 2.2.30. The scattering properties keep changing depending on the depth of the grassland as well as of the content of bushes and trees and, when it comes to forests, on the density and depth of the tree layer.

Figure 2.2.30
Schematic portrayal of the migration from Grassland (A and B) to Bush- (C) and Tree-Savannah (D). From Walter (1999)

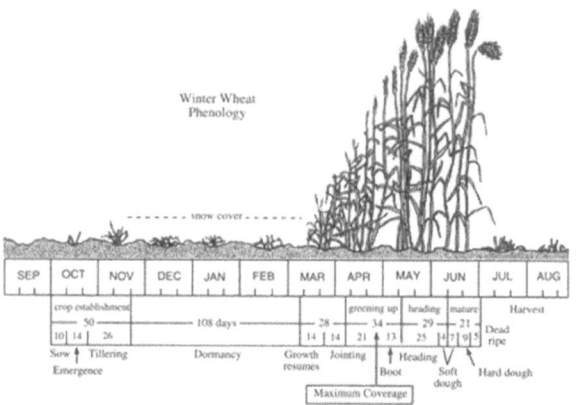

Large areas of land are used for agriculture. Grains grow in plants of the grass-family. The agricultural grasses are usually annuals, showing a distinct growing-pattern, alternating between un-vegetated and grass-covered ground. One such pattern is shown in Figure 2.2.31 for winter wheat in the Great Plains. This has significant influence on satellite-retrievals.

Figure 2.2.31
The phenological cycle of hard red winter wheat in the Great Plains of the United States. The crop is established in October and November. It lies dormant under snow cover until March, when growth resumes. The plants green up in April, produce heads in May, and mature in mid-June. The wheat is usually ripe and harvested by early July. Remotely sensed data acquired in October and November provide information on the amount of land prepared during the crop establishment period. Imagery acquired during the green-up phase in April and May can be used to extract information on standing-crop biomass and perhaps predict the harvested wheat yield. From Jensen (2000)

3 DOAS

Differential Optical Absorption Spectroscopy (DOAS) is a method for analysing the trace-gas composition of the atmosphere. For the method in general it is unimportant whether the sensors are placed on the ground, on a moving platform like a car, a ship or an aircraft or on a satellite. This does have effect on the particular retrievals though. This chapter introduces the general principle of DOAS and why we started using this method for remote sensing of vegetation.

Platt & Stutz (2008) explain that the basis of the early spectroscopic measurements, and many present quantitative trace gas analytical methods in the atmosphere and the laboratory, is Lambert – Beer's law, often also referred to as Bouguer – Lambert law. The law was presented in various forms by Pierre Bouguer in 1729, Johann Heinrich Lambert in 1760, and August Beer in 1852. Bouguer first described that, 'In a medium of uniform transparency the light remaining in a collimated beam is an exponential function of the length of the path in the medium'. However, there was some confusion in the naming of this law, which may be either the name of an individual discoverer or combinations of their names.
In this thesis I refer to it as the Lambert-Beer law.

Since the late 19th century, absorption spectroscopy is applied to the atmosphere. However, the concentration of the gas can be determined with the Lambert-Beer law, only if cross-sections are known accurately and all other factors can be eliminated, like e.g. stray light or detector-noise. This can be achieved in a laboratory, where all factors are controlled. It is generally not possible in the atmosphere due to scattering processes in the atmosphere. That is why Differential Optical Absorption Spectroscopy was developed, using the fact that aerosol extinction processes, instrument characteristics and the effect of many absorbers show very broad band structures. These factors become insignificant, since DOAS only evaluates narrow band structures.

Today DOAS is used for many applications and on many different platforms. Some are shown in Figure 3.1. For a more comprehensive introduction to different DOAS setups see Platt & Stutz (2008).

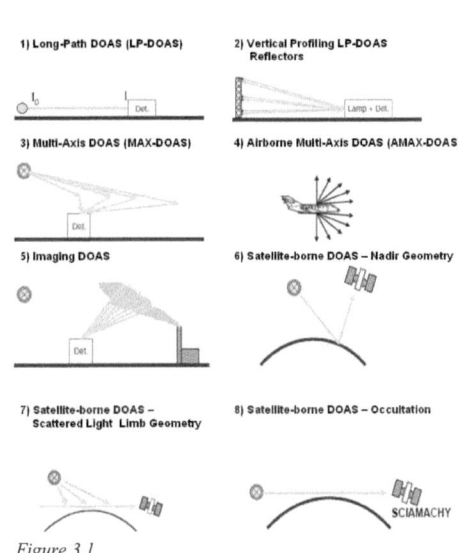

Figure 3.1
Different possible setups for DOAS. 1) and 2) are examples for active DOAS, which means that the light-source is artificial and the light-path is predetermined. In 1) the light-source is directed at the detector. In case 2) the light is sent out from the direction of the detector towards reflectors and then scattered back to the detector. When reflectors are placed apart, measurements of profiles or volumes become possible. 3) to 8) show setups for passive DOAS. This means that either direct or scattered natural light is used. An example for direct light measurements is case 8), where SCIAMACHY looks through the atmosphere directly at the sun. All other setups use scattered natural light. Adapted from Platt & Stutz (2008)

3.1 Method

The DOAS-method was originally developed to measure the gas content in the atmosphere. Gases absorb parts of the light when it is penetrating through them. If the light is later spectrally divided, the characteristic absorption structures can be seen. Every gas has its specific spectral regions where the light gets fully or partially absorbed. By identifying these absorption bands, the gases contained in the mixed environment can be identified.

3.1.1 Lambert-Beer law

DOAS uses the principle that gases absorb certain wavelengths of light passing through them. which is expressed in the Lambert-Beer law:

This law states that the intensity of light is reduced exponentially (depending on absorption cross- section, concentration and path length) when passing through an absorbing species.

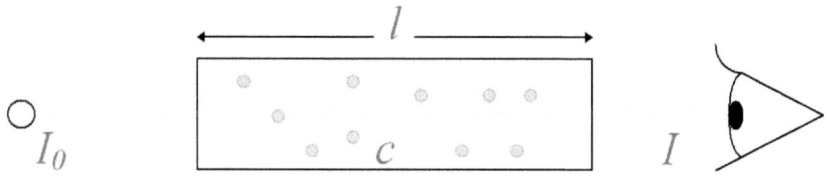

Figure 3.1.1:
Visualizing the parameters of the Lambert-Beer law: The intensity of light is reduced exponentially when passing through an absorbing species. From Wagner et al. (2008 b).

(3.1) $\quad I = I_0 \cdot \exp(-\sigma \cdot c \cdot l)$

- σ absorption cross -section of chemical species
- c concentration of absorbing chemical species
- l length of path through absorbing chemical species
- I intensity of Light at measuring device
- I_0 intensity of Light at source

In other words the higher the concentration of the gas, the more intensive is the absorption and the longer the light path through the gas, the more intensive is the absorption. The wavelength dependency of the absorption cross-section is characteristic to the chemical species.

3.1.2 Differential absorption

According to the Lambert-Beer law the composition of the light received at the sensor depends on the light source, on the concentration of the gas along the light path and on the specific cross- section of the gas which is depending on the wavelength.

The logarithm of the ration of equation (3.1) is called optical thickness $\tau(\lambda)$:

(3.2) $\tau(\lambda) = -\ln \dfrac{I(\lambda,\sigma)}{I_0(\lambda)}$

This can then be re-written as:

(3.3) $\tau(\lambda) = \sigma(\lambda,T) \int c(s)\,ds$

For easier readability the expression can be simplified by replacing $\int c(s)\,ds$ with S. The optical thickness is then defined as shown in equation 3.4.

(3.4) $\tau(\lambda) = \sigma(\lambda,T) S$

S describes the concentration of the absorbing gas integrated over the complete light path and is also called the Slant Column Density (SCD). When the cross-section of the gas $\sigma(\lambda, T)$ is known from laboratory measurements and the optical thickness can be calculated from I and I_0, then the gas concentration for the light path can be calculated if the gas concentration can be considered the same for the whole path.

Up to this point we have only considered a single gas responsible for absorption. However nature is much more complex and there are numerous gases and trace gases in the atmosphere. In addition the light is not only absorbed by the gases in the atmosphere but also gets scattered by molecules, aerosols and water droplets. Moreover there is an influence from the ground where the light is reflected. This was discussed in Chapter 2 of this thesis. All these effects can be combined in equation 3.5, where the effects of albedo and atmospheric scattering are expressed in the term $g(\lambda)$.

(3.5) $I(\lambda) = I_0(\lambda) \cdot g(\lambda) \cdot e^{-\sum \sigma_i(\lambda,T) S_i(\lambda)}$

After taking the logarithm, this equation can be regrouped to the following:

(3.6) $\ln \dfrac{I(\lambda)}{I_0(\lambda)} = -\sum \sigma_i(\lambda,T) S_i(\lambda) + \ln(g(\lambda))$

Unfortunately the equation cannot be solved in this way, since atmospheric scattering is not constant. Also the spectral reflectance from the ground is variable. To solve this problem, the differential optical absorption spectroscopy (DOAS)-method was developed. The basic principle of DOAS is that it uses the fact that the effects which are combined under the term $g(\lambda)$ are expressed as broad band structures in the spectrum, whereas the spectral absorption of gases studied with DOAS is very short band. The broad band effects can be filtered out by splitting up the cross-section $\sigma(\lambda)$ into fast-changing parts $\sigma'(\lambda)$ and in slow changing parts $\sigma_c(\lambda)$. This is expressed in equation 3.7.

(3.7) $\sigma(\lambda) = \sigma_c(\lambda) + \sigma'(\lambda)$

The difference between fast- and slow- changing spectral features depends on the wavelength range and the width of the expected absorption bands.

The fast-changing absorption $\sigma'(\lambda)$ is also called differential absorption cross section. Since DOAS is able to evaluate measurements for several atmospheric trace gases at the same time, it is often used for atmospheric measurements.

The higher the concentration of a gas, the more intense is its absorption at a characteristic wavelength. Only the depth of the absorption is analysed. Any changes in for example the intensity of the light source have an effect on the strength of the general signal, but not on the relative depth of the absorption-line. This is an elegant way to keep measurements relatively free of other influences apart from atmospheric composition.

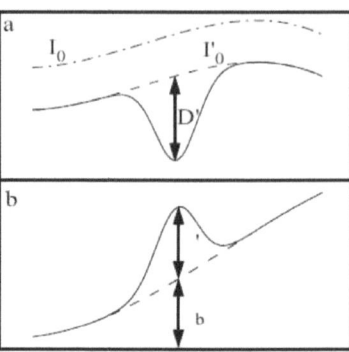

Figure 3.1.1
Principle of DOAS: I_{0} and σ are separated by an adequate filtering procedure into a narrow (D', and σ') and broad band (I'$_{0}$, and σ_{b}) part. From Stutz (1996)

This is of course simplified. Changes in the atmosphere due to scattering, Ring-effect, clouds and albedo alter the concentration of the gases and especially the length of the light path. Every absorbing chemical species needs to be extracted with its own cross-section. But which species are present when? And which light path can be assumed? For an detailed explanation see Platt and Stutz (2008). To give an impression of cross-sections for different trace-gas species in the atmosphere see Figure 3.2.1.

It is important to identify those areas where the signal of a chemical species can be clearly separated from other species (spectral overlap is no problem as long as patterns are distinguishable) while still being strong enough for separation from background noise.

The cross-sections are fitted together with a polynomial (to account for g (λ)) of a specified degree by means of a non-linear least squares fitting algorithm (see Gomer et al. (1993), Stutz & Platt (1996)).

3.2 Application to satellite data

The measurements at a satellite are affected by two different phenomena. In the first the satellite measures the reflected light which originates from the sun. The sun is composed of different elements and gases which produce their own absorptions. The visible absorption bands of the incoming light are called the Fraunhofer lines.

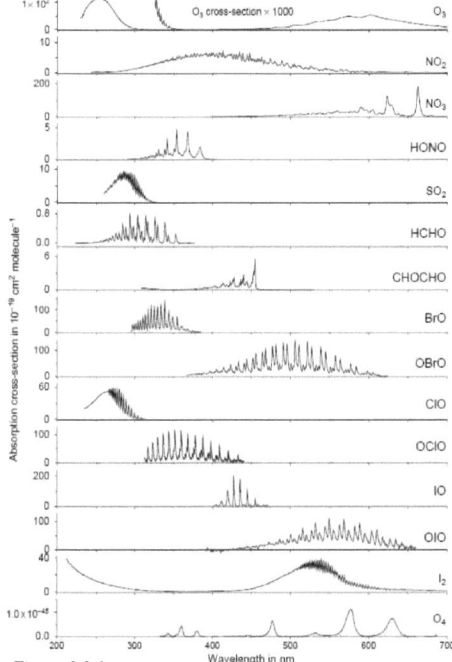

Figure 3.2.1
Details of the absorption cross-section features of a number of trace-gas species of atmospheric interest as a function of wavelength (in nm). Note the 'fingerprint' nature of the different spectra. From Platt & Stutz (2008)

In the second, before the light can be measured by the satellite, it has to pass through the atmosphere twice. Different gases absorb in their specific wavelength ranges and influence the outgoing light. Many gases in the atmosphere are variable over time.

The length of a satellite light path is significantly influenced by the scattering properties inside the atmosphere. Thus the length of the light path is very hard to determine. The concentration of the absorbing gas is highly variable over time but can be evaluated from the DOAS-equation. Several gases can be determined at the same time, so that the Lambert-Beer law is actually applied in parallel to several species.

Trace gases in the atmosphere are not the only effect influencing the spectral features. Light reaching the sensor at the satellite first went through the atmosphere, was reflected by the earth's surface and passed through the atmosphere again. These influences are considered in the following.

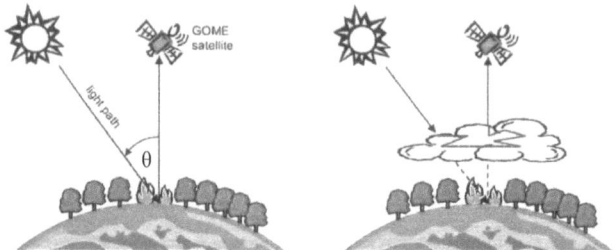

Figure 3.2.2
Scheme of the nadir viewing satellite-geometry (from Leue, 1991). In a cloud-free scene (left), the light-path crosses the troposphere and is reflected to the satellite with rel. simple geometry. The situation is more complex for a clouded scene (right), where the troposphere is partly or fully shielded while the light-path inside the cloud is lengthened. The air mass factor depends on clouds, aerosols and ground albedo.

With the DOAS method the gas concentration in the atmosphere is determined, integrated along the light path. This is called the Slant Column Density (SCD). The SCD is greatly influenced by clouds, aerosols, ground albedo and the processes in the atmosphere. To derive a quantitative result for trace-gas concentrations, the SCD is usually converted to Vertical Column Densities (VCD) by using the Air Mass Factor (AMF):

$$(3.8) \quad VCD = \frac{SCD}{AMF}$$

Figure 3.X schematically shows the geometry of nadir satellite observations. The light-path and with it the AMF depends strongly on the Solar Zenith Angle θ (SZA). See also Figure 3.2.2, for a simple reflection of the sunlight from the ground the "geometric AMF" can be applied:

$$(3.9) \quad AMF_{geom} = 1 + \frac{1}{\cos(\theta)}$$

Equation (3.9) is a good approximation for the conversion of SCD to VCD for stratospheric absorbers and SZAs below about 70°. "For trace gases in the troposphere the determination of AMF is more complex. The actual light-path is determined by ground albedo, aerosols and clouds. For a cloud-free scene with high ground albedo, the light is predominantly reflected at the ground. For dark surfaces (or high aerosol load) the role of light scattered in the atmosphere increases. For a clouded scene, finally, the air masses below the cloud may be completely shielded and are "invisible" for the satellite." (Beirle, 2004) For atmospheric gases, the AMF also depends on the gas profile. The AMF for a trace gas above the cloud-layer is different from an AMF for a trace gas underneath the cloud-layer, even if all other factors remain the same.

Another influence on the light reaching the satellite occurs when the light is reflected from the ground. This is the aspect with which this thesis is most concerned.

The length of a satellite light path is intensively influenced by the scattering properties inside the atmosphere. It may also be influenced by multiple scattering inside a vegetation canopy (see also Figure 2.2.20). Thus the length of the light path is very hard to determine. The concentration of the absorbing gas or the absorbing pigment is highly variable in location and concentration but can be evaluated from the DOAS-equation.

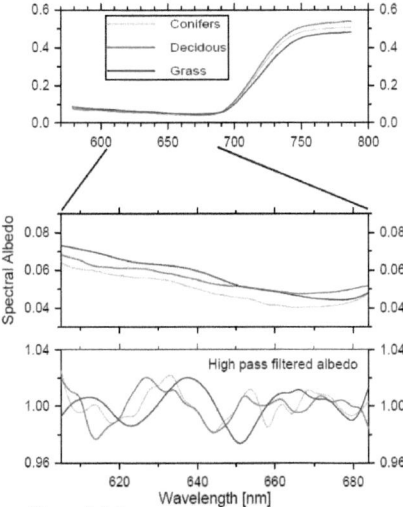

Figure 3.2.3
Top: Spectra of the reflectance over different kinds of vegetation, reproduced from the ASTER Spectral Library through the courtesy of the Jet Propulsion Laboratory. The strong change of the reflectance between the red and infrared part of the spectrum is usually exploited for the remote sensing of vegetation. In the red part of the spectrum the reflectance is small (middle), but contains characteristic spectral structures (displayed after high-pass filtering, bottom). From Wagner et al. (2007).

Albedo influences not only the general strength of the signal, but also the fine spectral signals used for DOAS. This was demonstrated by Wagner et al. (2007), when they fitted high-pass filtered vegetation spectra analogue to a cross-section. See Figure 3.2.3. The applied vegetation spectra were from the ASTER Spectral Library through the courtesy of the Jet Propulsion Laboratory, California Institute of Technology, Pasadena, California. © 1999. California Institute of Technology.

This confirmed that vegetation did influence the high-frequency part of the measured spectrum. This contradicts the original DOAS-assumption: that reflection properties on the ground influence only the slow-changing part of a cross-section (expressed in the term $g(\lambda)$ in equations 3.5 and 3.6).

To extract vegetation information with DOAS, reference spectra in similar fashion and spectral resolution to the trace gas spectra are needed. Those did not exist before, which is why they had to be generated first. The measurement procedure is described in detail in Chapter 6.

4 Satellite Instruments and Data

There has been a succession of satellites being able to provide data for the DOAS-method. The first of the satellites was launched in April 1995, the GOME-instrument, followed by the SCIAMACHY-instruments and that again followed by the GOME2-instrument. All three satellite-instruments are operational right now although the older sensors have certain limitations. We also use other data sets as a basis for our retrievals and for comparison of our results. For ground truth comparison of vegetation we used the Globcover map; produced by ESA on basis of the MERIS-instrument. All these instruments and data sets are introduced in this chapter.

The DOAS-instruments produce a continuous spectrum with moderate spectral resolution while MERIS measures in 15 distinct spectral channels with lower spectral resolution. MERIS also produces medium spatial resolution, while the DOAS-instruments only provide low spatial resolution.

4.1 GOME

The first instrument we used was the GOME-instrument which is flying on the ERS-2 satellite. GOME is the abbreviation of Global Ozone Monitoring Experiment. This gives the first introduction to the original dedication of the instruments which was to monitor the ozone hole. The other DOAS-applications were developed later, giving the instrument a much broader use.

The GOME-instrument is flown on board of the ERS-2 satellite. This satellite was launched on 21 April 1995. The first data used in this thesis originated from measurements in 1996. The ERS-2 satellite is still in orbit and still operational. This is also true for the GOME-instrument, although there certain limitations apply due to aging. This is why we restricted our use of the GOME-data from 1996 to 2004.

The satellite operates in a near polar sun-synchronous orbit at an altitude of 780 km with an equator-crossing-time of approximately 10.30 am local time. While the satellite orbits the earth on an almost north-south direction, the GOME-instruments scans the surface of the earth

Figure 4.1.1
An artists impression of ERS-2 in flight. © ESA

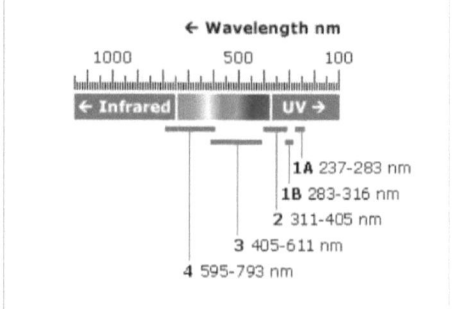

Figure 4.1.2
GOME spectral channels. From AT2-ELS.

42

from nadir in the perpendicular east-west direction.

GOME is measuring in four different channels at wavelength ranges from 237 to 793 nm. The spectral resolution of the different channels varies between 0.2 nm and 0.4 nm. The division into channels is shown by Figure 4.1.2. The ground resolution of the pixels is 320 km east-west to 40 km north and south. One scan from east-west contains three scans side by side. So the swath-width of the satellite is 960 km. In this fashion the satellite needs three days to cover the whole surface of earth.

GOME-data has a very characteristic area of missing data, roughly over Central Asia. For the daily reference measurement of the sun during this period in flight the satellite instrument actually measures directly into the sun to get the daily sun reference.

4.2 SCIAMACHY

On the night of 28 February/1 March 2002 the European Space Agency (ESA) launched the ENVIronmental SATellite (ENVISAT) on an Ariane 5 rocket from the European spaceport in Kourou.

ENVISATs mission is to collect information on environmental parameters on previously unknown levels of quality and inter-comparability. Figure 4.2.1 Shows a scetch of the sattelite with attached instruments.

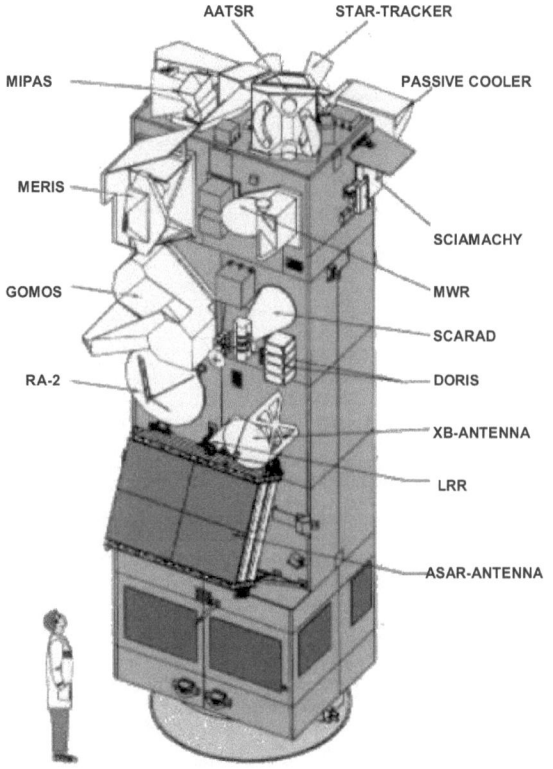

Figure 4.2.1: The European research satellite ENVISAT. All instruments are shown and named. (©ESA).

SCIAMACHY was started in 2002 und is still operational. However, the data recorded after October 2010 needs special care when compared to older data: Until recently, ENVISAT has been operating in a near polar, sun-synchronous orbit at an altitude of about 800 km. Since The instruments aboard ENVISAT far outlasted their expected lifetime, the satellite was brought to a new orbit with several manoeuvres between 22 October and 2 November 2010. The new orbit is not sun-synchronous anymore. Data of ENVISAT instruments after the orbit shift is not considered in this thesis.

Unlike GOME, the SCIAMACHY instrument alternates between limb and nadir measurement modes. See also figure 4.2.2. As a third mode, sun/moon occultations are possible.

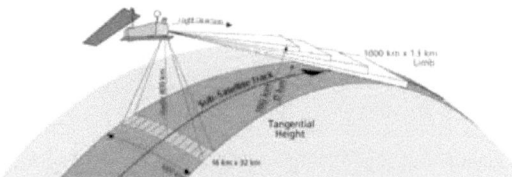

Figure 4.2.2
Sketch of possible scan modes of SCIAMACHY. © ESA

In nadir mode a swath-width of 960 km gives full global coverage every six days (see Fig. 4.2.3. for an example of an orbit and the swath). The typical ground pixel size of SCIAMACHY is 30 km (approx. north-south) times 60 to 120 km (approx. east-west, depending on resolution-mode). In this thesis only nadir measurements are used, since only this mode registers reflection of vegetation on the ground.

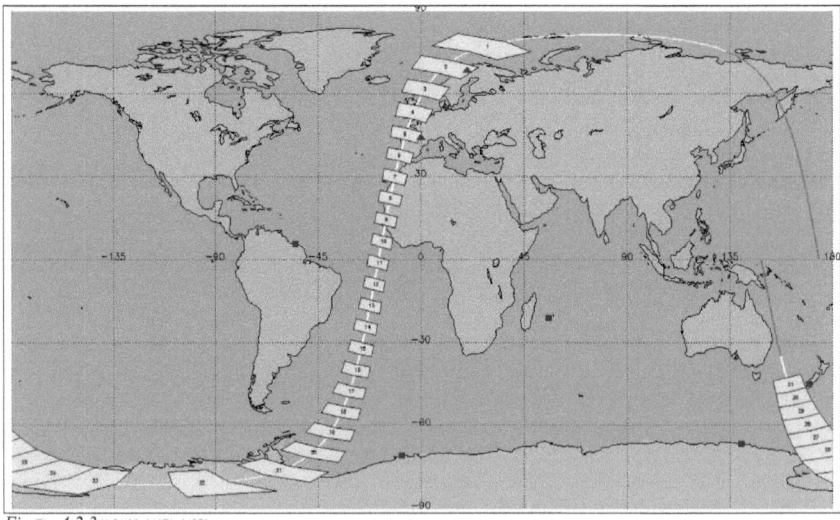

Figure 4.2.3
Example of SCIAMACHY nadir orbit ground track (taken from the Sciamachy Operations Support Team (SOST) website: http://atmos.af.op.dlr.de/projects/scops/)

SCIAMACHY (SCanning Imaging Absorption spectroMeter for Atmospheric CHartographY) is a grating spectrometer, measuring solar radiation reflected at the earth's surface or scattered within the atmosphere.

SCIAMACHY is a superset of its predecessor GOME (Global Ozone Monitoring Experiment) onboard ERS-2.

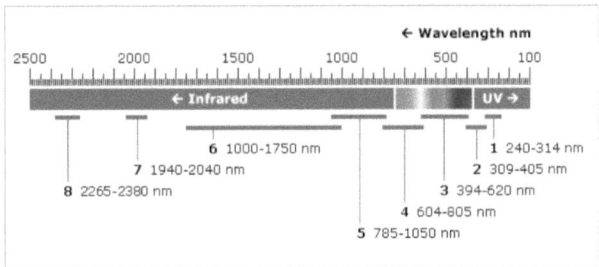

Figure 4.2.4
SCIAMACHY spectrometer channels. From AT2-ELS

SCIAMACHY measures in 8 spectral channels, providing continuous spectra from 240 to 1750 nm plus two additional spectral ranges in the IR. The resolution is between 0.2 nm and 1.5 nm, depending on spectral channel. Figure 4.2.4 shows the spectral ranges of the different channels.

The different channels have different spectral resolutions. Figure 4.2.5 provides the necessary information.

Instrument Parameters			
	Channel	Spectral Range	Spectral Resolution
High Resolution Channels	1	240-314 nm	0.24 nm
	2	309-405 nm	0.26 nm
	3	394-620 nm	0.44 nm
	4	604-805 nm	0.49 nm
	5	785-1050 nm	0.54 nm
	6	1000-1750 nm	1.48 nm
	7	1940-2040 nm	0.22 nm
	8	2265-2380 nm	0.26 nm
Polarisation Measurement Devices (broadband)	PMD 1 to 7	310-2380 nm	67 to 137 nm (channel dependent)
Altitude Range		10 km -100 km depending on measurement mode	
Vertical Resolution		2.4 km - 3 km depending on measurement mode	
Operation		continuously over full orbit	
Data Rate		400 kb/s nominal, 1867 kb/s real time mode	
Mass		198 kg	
Power		122 W	

Figure 4.2.5
Instrument parameters for SCIAMACHY. © ESA

4.3 MERIS

The MEdium Resolution Imaging Spectrometer (MERIS) is an instrument aboard ENVISAT. The characteristics of this satellite were presented before in chapter 4.4.2.

MERIS is a programmable, medium-spectral resolution, imaging spectrometer operating in the solar reflective spectral range. Fifteen spectral bands can be selected by ground command, each of which has a programmable width and a programmable location in the 390 nm to 1040 nm spectral range.

The instrument scans the Earth's surface by the so called 'push broom' method. CCDs arrays provide spatial sampling in the across track direction, while the satellite's motion provides scanning in the along-track direction.

MERIS is designed so that it can acquire data over the Earth whenever illumination conditions are suitable. The instrument's 68.5° field of view around nadir covers a swath width of 1150 km. The Earth is imaged with a spatial resolution of 300 m (at nadir). This resolution is usually reduced to 1200 m by the on board combination of four adjacent samples across track over four successive lines. For the Globcover map only pixels at full resolution were used.

The 15 spectral bands of MERIS are as follows (Source: ESA):

Band Nr.	Band centre (nm)	Bandwidth (nm)	Potential Applications
1	412.5	10	Yellow substance and detrital pigments
2	442.5	10	Chlorophyll absorption maximum
3	490	10	Chlorophyll and other pigments
4	510	10	Suspended sediment, red tides
5	560	10	Chlorophyll absorption minimum
6	620	10	Suspended sediment
7	665	10	Chlorophyll absorption and fluo. reference
8	681.25	7.5	Chlorophyll fluorescence peak
9	708.75	10	Fluo. Reference, atmospheric corrections
10	753.75	7.5	Vegetation, cloud
11	760.625	3.75	Oxygen absorption R-branch
12	778.75	15	Atmosphere corrections
13	865	20	Vegetation, water vapour reference
14	885	10	Atmosphere corrections
15	900	10	Water vapour, land

4.3.1 Globcover and Vegetation-classes

Globcover v.2 uses MERIS full resolution mode data (300 m) acquired between April 2005 and May 2006. The thematic legend is compatible with the UN-land cover classification system (LCCS). This new product updates and complements the other existing comparable global products, such as the Global land cover map at 1 km resolution for the year 2000 (GLC2000) produced by the Joint Research Centre (JRC). It is expected to improve such previous global products in particular because of the finer spatial resolution.

The Globcover-product is based on ENVISATs MEdium Resolution Imaging Spectrometer (MERIS) level 1b-data acquired in full resolution mode with a spatial resolution of 300m. For the generation of the level 1b-data, the raw data have been resampled on a path-oriented grid, with pixel-values having been calibrated to match the top of atmosphere (TOA) radiance. For a detailed description of the classification process see GLOBCOVER Products Description and Validation Report (Bicheron et al. 2008).

Figure 4.3.1. shows a small version of the final result. For the thesis I use the digital version of the map. Source Data: © ESA / ESA GlobCover Project, led by MEDIAS-France. The Globcover map can be downloaded from http://ionia1.esrin.esa.int/

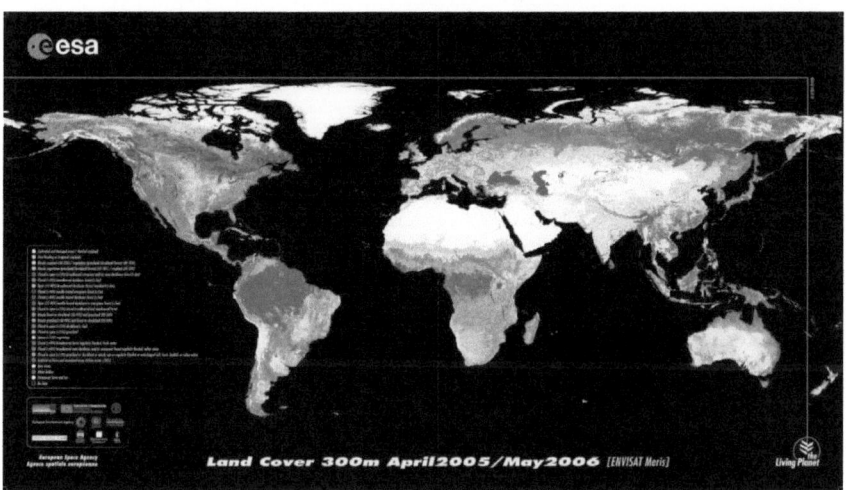

Figure 4.3.1
Overview of Globcover map. The digital version has a ground resolution of 300 by 300 m^2. © ESA

The vegetation-classes used in Globcover are compatible with other global products like the UN-land cover classification system. But they are not directly translatable to the measurements in this thesis. This classification differentiates between deciduous or broadleaved forest, coniferous forest and grasslands of different heights and per cent land cover. The measurements in this thesis differentiate more between coniferous tree species, broadleaved tree species and different grass lands while not considering height or percent land-cover. So the comparison is not straight forward. A relation of Globcover-classes to our own measurements is given in chapter 7.

Figure 4.3.2 shows an enlargement of figure 4.3.1, rendering the legend readable.

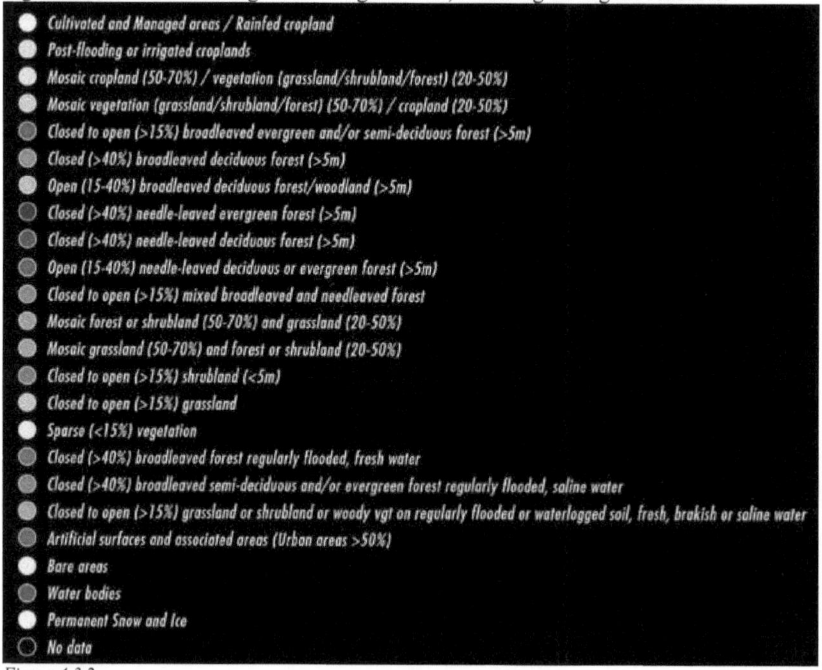

- Cultivated and Managed areas / Rainfed cropland
- Post-flooding or irrigated croplands
- Mosaic cropland (50-70%) / vegetation (grassland/shrubland/forest) (20-50%)
- Mosaic vegetation (grassland/shrubland/forest) (50-70%) / cropland (20-50%)
- Closed to open (>15%) broadleaved evergreen and/or semi-deciduous forest (>5m)
- Closed (>40%) broadleaved deciduous forest (>5m)
- Open (15-40%) broadleaved deciduous forest/woodland (>5m)
- Closed (>40%) needle-leaved evergreen forest (>5m)
- Closed (>40%) needle-leaved deciduous forest (>5m)
- Open (15-40%) needle-leaved deciduous or evergreen forest (>5m)
- Closed to open (>15%) mixed broadleaved and needleleaved forest
- Mosaic forest or shrubland (50-70%) and grassland (20-50%)
- Mosaic grassland (50-70%) and forest or shrubland (20-50%)
- Closed to open (>15%) shrubland (<5m)
- Closed to open (>15%) grassland
- Sparse (<15%) vegetation
- Closed (>40%) broadleaved forest regularly flooded, fresh water
- Closed (>40%) broadleaved semi-deciduous and/or evergreen forest regularly flooded, saline water
- Closed to open (>15%) grassland or shrubland or woody vgt on regularly flooded or waterlogged soil, fresh, brakish or saline water
- Artificial surfaces and associated areas (Urban areas >50%)
- Bare areas
- Water bodies
- Permanent Snow and Ice
- No data

Figure 4.3.2
Legend of Globcover. © ESA

4.4 HICRU

The Heidelberg Iterative Cloud Retrieval Utilities (HICRU) algorithm retrieves effective cloud fraction, cloud height and surface albedo. Reflectances from broad-band spectrometers (617-705nm) and spectral analysis (DOAS) of the O_2-γ-band around 630nm are used. The included radiative transfer uses the Monte-Carlo models TRACY II and McArtim. The algorithm is applied to SCIAMACHY and GOME. For a more comprehensive description see Gzegorski (2009).

HICRU was used for both GOME and SCIAMACHY data to choose suitable low-cloud pixels.

5 Mini-MAX-DOAS

Mini-MAX-DOAS is the abbreviation for a mini (small) Multi-Axis Differential Optical Absorption Spectrometer. Multi-Axis capability allows the instrument to rotate along its axis to programmed angles. A Mini-MAX-DOAS also includes a cooling device to control the temperature inside the instrument, keeping the spectrometer at a constant predefined temperature. For this thesis I used an OceanOptics USB2000+ spectrometer.

5.1 Build-up of instrument

The core of our Mini-MAX-DOAS instrument is an OceanOptics USB2000+ spectrometer. It has a slit of 25 μm, grating #6 with a nominal wavelength range of 500 - 790 nm and a nominal resolution of 0.5 nm.

Figure 5.1.1 is a graphic from OceanOptics, showing the light path inside the USB2000+ and its components. Number (1) in the picture shows the input for the fibre which brings the light to the optical bench. (2) denotes the slit in the instrument, through which the light passes. (3) shows the first filter where light is restricted to

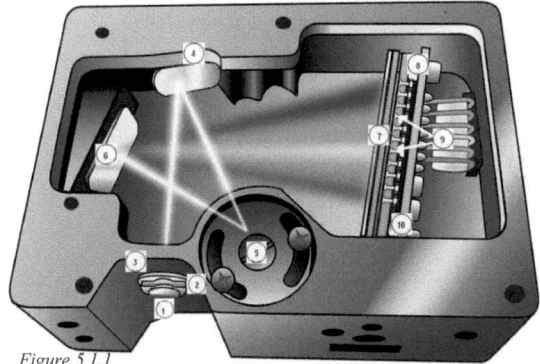

Figure 5.1.1
USB2000+ Spectrometer with Components. 1) connector for imput-fiber, 2) slit, 3) filter, 4) Collimating mirror, 5) grating, 6) focusing mirror, 7) detector collection lens, 8) detector. See USB2000+ Components Table for an explanation of the function of each numbered component in the USB2000+ Spectrometer in this diagram. From OceanOptics (2007)

the wavelength of interest for the spectrometer inside. (4) is the collimating mirror which sends the light towards the grating. The grating (5) defracts the light into its different wavelengths and sends it to the mirror (6) which focusses the light onto detector collection lenses (7), which concentrate the signal for the detectors (8). All these different features are also described in more detail in Table 5.1, provided by OceanOptics. The optical bench has no moving parts that can wear or break; all the components are fixed in place at the time of manufacture.

USB2000+ Components Table

Item	Name	Description
1	SMA 905 Connector	Secures the input fibre to the spectrometer. Light from the input fibre enters the optical bench through this connector.
2	Slit	A dark piece of material containing a rectangular aperture, which is mounted directly behind the SMA connector. The size of the aperture (25 μm) regulates the amount of light that enters the optical bench and controls spectral resolution.
3	Filter	Restricts optical radiation to pre-determined wavelength regions. Light passes through the filter before entering the optical bench.
4	Collimating Mirror	A SAG+, Ag-coated mirror focuses light entering the optical bench towards the grating of the spectrometer. Light enters the spectrometer, passes through the SMA connector, slit, and filter, and then

		reflects off the collimating mirror onto the grating.
5	Grating	A #6 (1200 lines per millimeter, blazed at 750 nm) grating diffracts light from the collimating mirror and directs the diffracted light onto the focusing mirror.
6	Focusing Mirror	A SAG+, Ag-coated mirror receives light reflected from the grating and focuses first-order spectra onto the detector plane.
7	Detector Collection Lens	Attaches to the detector to increase light-collection efficiency. It focuses light from a tall slit onto the shorter detector elements. The detector collection lens should be used with large diameter slits or in applications with low light levels. It also improves efficiency by reducing the effects of stray light.
8	Detector	Collects the light received from the focusing mirror or detector collection lens and converts the optical signal to a digital signal. Each pixel on the detector responds to the wavelength of light that strikes it, creating a digital response. The spectrometer then transmits the digital signal to the Spectra Suite application.

Table 5.1
Components of the USB2000+. Adapted from OceanOptics (2007)

Figure 5.1.2 shows the instrument used for this thesis with all the cables connected. The cables (from left to right) are: control of the multi-axes motor, connection for OceanOptics device to the computer, connection for the temperature and motor control to the computer and to the far right connection to the 12V-battery. The ribs and fan on top are part of the cooling device. The telescope is attached to limit stray light interfering with the measurements. The Multi-Axis motor is on the opposite side of the instrument.

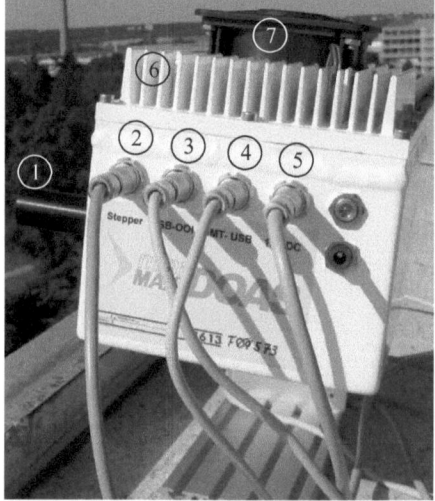

Figure 5.1.2
The Mini-MAX-DOAS instrument used for this thesis.
1) telescope, 2) cable to axis-motor, 3) cable controlling spectrometer, 4) cable controlling internaö sensors and cooling, 5) cable to 12V battery, 6) cooling-ribs, 7) fan

5.2 External sources of interference on measurements

A Mini-MAX-DOAS is very sensitive to external influence. It has to be calibrated with great care considering spectral calibration, temperature and light source. The reasons and the methods are introduced in this chapter.

5.2.1 Temperature

Measurements with a Mini-MAX-DOAS have to be done under very stable temperature conditions inside the spectrometer. To keep dark current as low and stable as possible the best operating temperature for our specific Mini-MAX-DOAS had to be evaluated. To derive that, spectral resolution at different temperatures was compared, measuring the distinct peaks of the white computer screen. Usually these measurements are done with a calibration lamp. The computer screen was chosen since it also works with discharge tubes and was an easily accessible highly structured light source. Figure 5.2.1 demonstrates the results for 20, 10 and 0 degrees Celsius. Please note the loss of resolution, which is best to be seen inside the red circle. Since these losses in resolution already appear well above 10°C, we had to decide for a minimum temperature of 16°C to avoid too much loss in resolution. Although a temperature of 0°C would be more desirable to keep dark current low, the loss of resolution was not acceptable. Therefore all measurements were done at an internal temperature of 16.8°C.

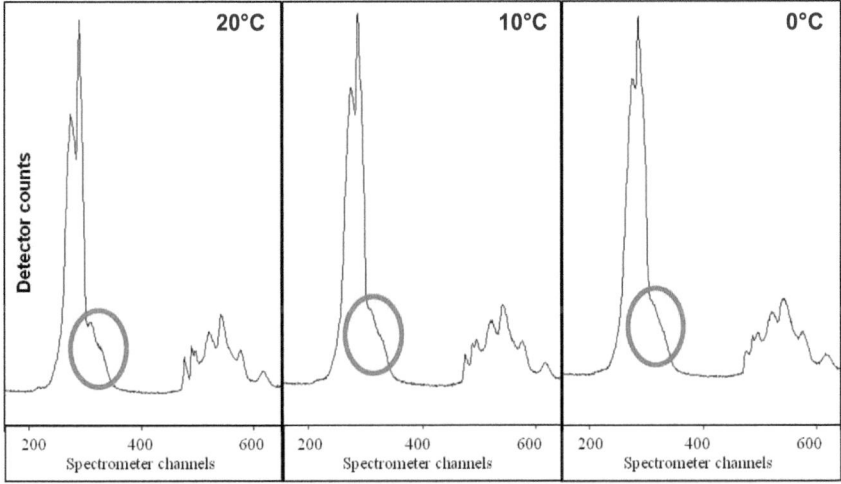

Figure 5.2.1
The effect of spectrometer-temperature on resolution of measurement. The Mini-MAX-DOAS measured the radiation of the white computer screen. The red circle highlights one example of diminishing resolution with decreasing temperature. The sub-peaks visible at 20°C merely show as shoulders at 10°C and disappear completely at 0°C.

The instrument needs time to reach the optimum working temperature. After the temperature is set, the instrument also needs time to cool the interior of the Mini-MAX-DOAS. Stable temperature inside the spectrometer are recognisable by a steady output of similar measurement values when pointed at an unchanging object. This process is demonstrated in Figure 5.2.2, where the dark blue line describes the internal temperature of the instrument and the pink and yellow lines denote the maximum and average measurement values. The temperature reaches its plateau much sooner than the detector counts. The instrument behaviour suggests a minimum of 30 minutes' delay between setting the instrument temperature and the onset of measurements.

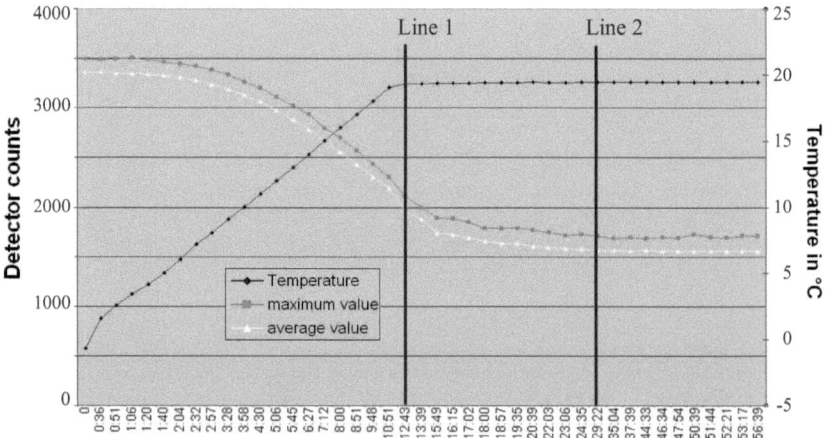

Figure 5.2.2
Time-lag between setting internal temperature, stable internal temperature and stable measurement values – defining the moment of stable temperature inside the spectrometer. Line 1 shows the time required to reach a stable temperature inside the Mini-MAX-DOAS. Line 2 shows the time when detector counts stabilize, indicating a stable temperature of the spectrometer.

5.2.2 Light-source

To produce reliable reference data the influence of the light source on the measured data has to be known. To correct for changes of the incoming light, the reflection of the light source over granulated salt was measured repeatedly as reference. Granulated salt reflects very effectively and causes no absorption of it's own within the studied wavelength-range.

For our first measurements, natural day light was the light-source. To reduce the possible influences, measurements were restricted to absolutely cloud-free days. On those days measurements were done only around solar noon to have only small influence through the changing solar zenith angle. Also, vegetation and reference were measured as close together in time as possible. Sometimes the delay was only a very few seconds. However this was not fast enough on most days to avoid atmospheric interference. The processes discussed in Chapter 2.1 obviously change rapidly in nature.

The strongest atmospheric signals in the measurement range of 500-800 nm are water-vapour, O_2 and – indirectly via the Ring-effect - Fraunhofer lines. The amount of water-vapour is a highly variable component in the atmosphere, with rapid spatial and temporal changes. Still, the apparent influence of water-vapour in our reference measurements came as a surprise since those days apparently had absolutely blue sky conditions.

Comparisons with measurements of vegetation under artificial light show the atmospheric effects to the measured vegetation reflectance. This is demonstrated in Figure 5.2.3. In A) measurements over Zea mays are shown. The measurement under artificial light is shown blue; the natural light-measurement is shown in pink. Comparisons to the reference spectra for atmospheric absorbers in B) demonstrate the influence of atmospheric absorbers on the vegetation measurements. Three examples are highlighted by lines. The blue line shows an obvious influence of the Fraunhofer-spectrum (caused by changes in Raman-scattering), the pink line shows the influence of water vapour and the green line that of oxygen. The results make it obvious that for new vegetation reference spectra an artificial light source must be chosen. The artificial light-source is described in Chapter 6.1.2.

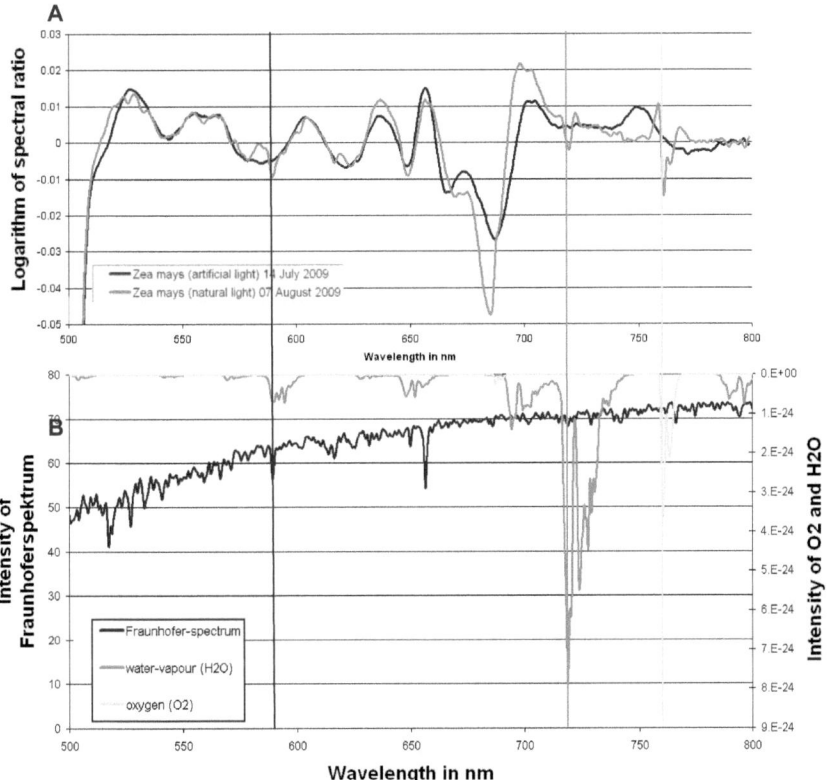

Figure 5.2.3
Comparison of vegetation reflectance spectra measured with natural or artificial light source. A) shows measurements over Zea mays (maize) with natural daylight (pink line) and artificial light (blue line) as light-source. The pink line shows many high-frequency structures that are not apparent in the blue line. B) displays the reference spectra (source: HITRAN database) used in this thesis. Comparing A) to these reference-spectra of Fraunhofer lines (dark blue), water-vapour (pink) and O_2 (green) explain the sources of those additional structures. For easier comparison three coloured lines are added, connecting apparent assocoations. The blue line connects the natural light peak over Zea mays with the corresponding peak in the Fraunhofer-spectrum: the pink line does the same for water-vapour. The green line connects the oxygen-reference to the natural light measurement.

5.3 Spectral calibration

The Mini-MAX-DOAS has to be carefully calibrated. All channels of the spectrometer have to be assigned its specific wavelength in nm. To achieve this, a measurement of atmospheric light is compared to reference spectra for atmospheric absorbers. (Source: HITRAN database)

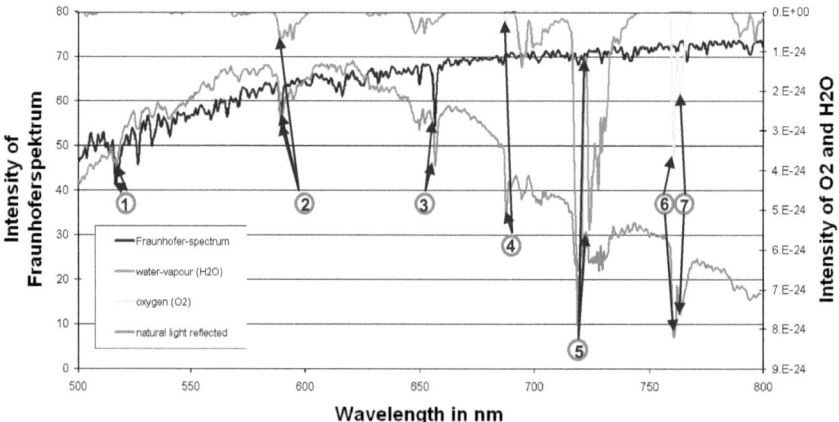

Figure 5.3.1
Connecting the spectral structures of natural-light-measurement to reference spectra (source: HITRAN database). The information in nm from reference and spectral channel from MiniMax DOAS are given in Table 5.1.

Figure 5.3.1 displays the natural light spectrum in the same graph with the spectra of the sun and the atmospheric absorbers. Thus the relationships of the different peaks are easily apparent. We chose peaks across all the measurement range to ensure a good calibration.

The calibration points were then entered into a table and a second order regression was performed. The resulting equation is displayed in Figure 5.3.2. With this regression all other channels were calibrated to their own wavelength.

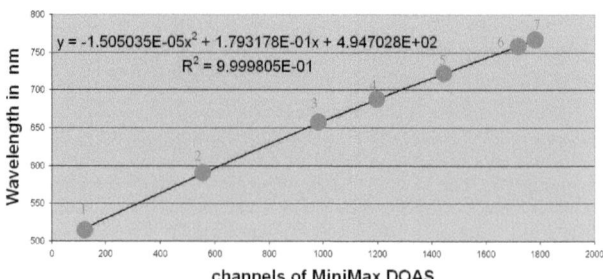

Figure 5.3.2
Resulting regression for wavelength calibration.

5.4 Spectral Resolution

To determine the spectral resolution of the MiniMax DOAS, we measured the reflected light of a red laser-pointer which was emitting near the centre of the measurement range. The FWHM (full width at half maximum) is 0.95 nm.

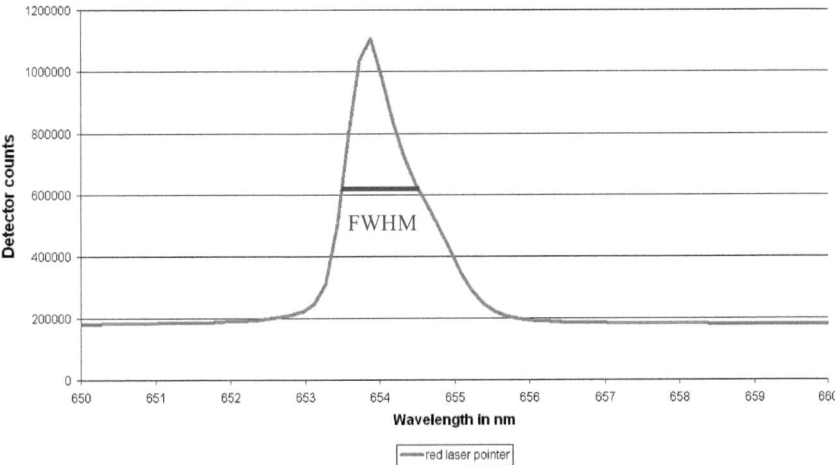

Figure 5.4.1
Reflection of Laserpointer over Salt. The FWHM is 0.95 nm in the red spectral range.

5.5 Spectral stability

To monitor the spectral stability of the MiniMax DOAS, we recorded continuously reflectances of natural light. Some results of the comparisons are shown in Table 5.2. The first column gives the reference points introduced in chapter 5.3; the second column, denoted MiniMax, gives the instrument-channels, to which we first referenced them in 2008, followed by the correlated wavelength.

In this section we investigate the spectral features at four different points in time, twice each in 2009 and 2010. The second to last column gives the wavelength in nm correlated with the last measurements (on 04 October 2010). The last column gives difference in nm between the first and the last correlation. All shifts are well below the spectral resolution of the MiniMax DOAS.

Almost all vegetation-measurements were done in 2009, when spectral shift was still negligible.

	MiniMAX channel	nm	01.04.09	18.11.09	21.06.10	04.10.10	nm	nm difference
1	127	517.2334	126	126	126	126	517.0579	0.1755
2	555	589.5883	555	553	553	553	589.263	0.3253
3	984	656.5789	984	983	982	982	656.2795	0.2994
4	1196	687.6386	1196	1195	1194	1194	687.3519	0.2867
5	1444	722.2557	1444	1444	1443	1443	722.1198	0.1359
6	1739	761.0224	1739	1737	1738	1737	760.7684	0.254
7	1760	763.6822	1760	1759	1758	1757	763.303	0.3792

Table 5.2: Comparison of position of spectral feat

6 Measurement of Vegetation Reference Spectra

6.1 Methods of Measurement

The measurements had to be produced with particular care since they are supposed to serve as references for future satellite data-analysis. Therefore possible external sources of interference must be investigated since they might result in interference with the vegetation signal.

6.1.1 Potential Sources of Interference

As described in Chapter 2, many processes in the atmosphere influence the sunlight passing through. Most of these scattering or absorbing processes change over time, sometimes over very short periods of time.

For reasons described in more detail in Chapter 6.2.1, we divide the measured spectra over vegetation by measured spectra of reflected light over salt. The underlying assumption is that any spectral properties of the instrument, the light source or the atmosphere cancel out in this ratio. Since these measurements must be done by the same instrument, both measurements have a short time difference. Although this was often just a few seconds, it gave time enough for changes to occur in the atmospheric processes or the distribution of trace gases. Particularly changes in Raman-scattering and the absorption of water-vapour and oxygen became visible within the measurement-range. Changes in Raman-scattering were made especially apparent by Fraunhofer-lines overlaying the vegetation spectra. This was already discussed in Chapter 5.2.2 and is demonstrated in Figure 5.2.3.

Water vapour is an atmospheric component which undergoes rapid spatial and temporal changes. The potential water-content of the atmosphere is directly related to the temperature while the actual content also relies on possible sources of evaporation (e.g. open water, vegetation, dry rock etc.), mixing and chemical processes in the atmosphere. It can also switch phase from vapour to water and ice and back again.

Oxygen is one of the major components of the atmosphere, contributing 21% of the volume. It's absorption changes due to changing light-paths caused by e.g. Aerosols. Another possible interference results from a changing solar zenith angle (SZA) during measurements. For this reason the time for measurements was restricted to the time around noon, when the SZA changes at the slowest rate.

6.1.2 Requirements for measurements without external Interference

All the effects mentioned in Chapter 6.1.1 make it obvious that measurements with natural light are not feasible for producing reference spectra over vegetation. An artificial light-source had to be used while blocking out sunlight at the same time.

The artificial light-source for our reference-measurements has to produce a spectrally continuous illumination within the whole measurement range. This immediately disqualifies

LEDs, since they radiate only at very specific spectral ranges and only marginally in the near infrared (see also Figure 6.1.1)

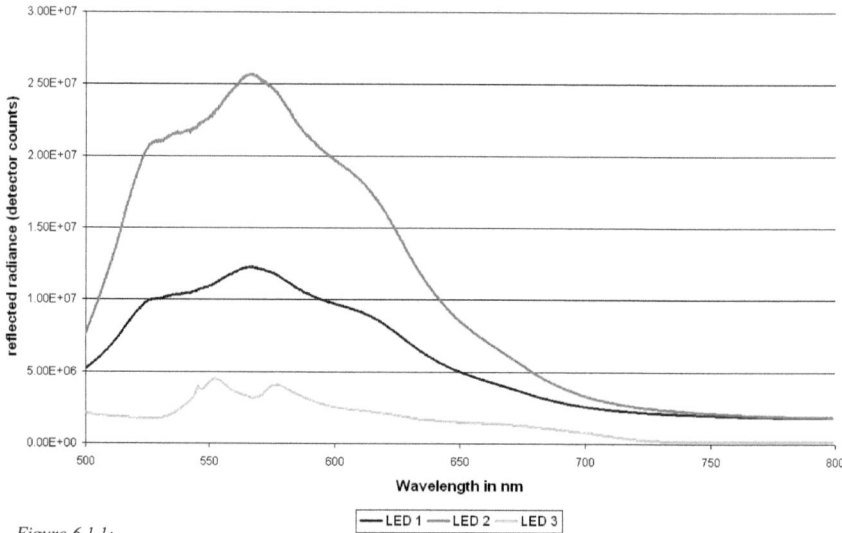

Figure 6.1.1:
Measured spectra of LEDs reflected from salt..

The radiation of Halogen lamps proved to be more suitable for our measurements (see Figure 6.1.2).

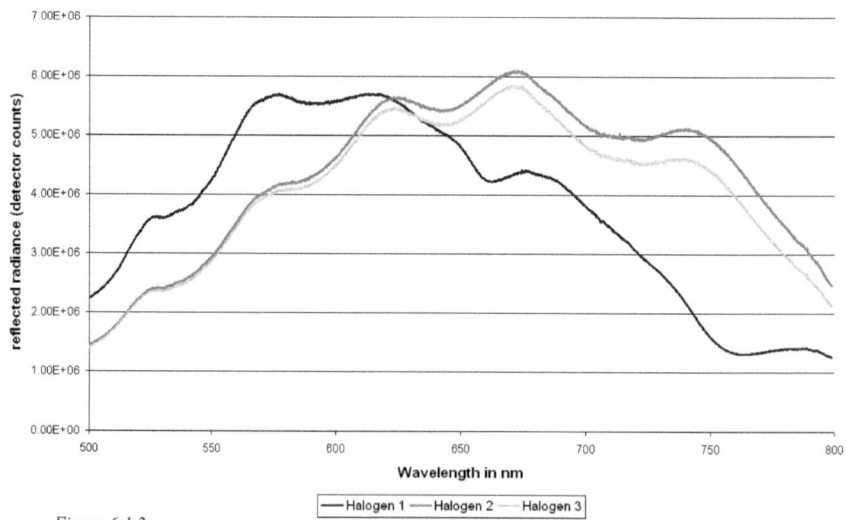

Figure 6.1.2:
Measured spectra of different Halogen-Lamps. Halogen 1 is less suitable for our measurements but Halogen 2 and 3 are good artificial light-sources.

To be able to use an artificial light-source during daytime, when vegetation is photosynthetically active, we had to exclude natural sunlight from our measurements. This could be done either inside a building, bringing fresh leaves from the botanical garden or – for the grasses – had to be done under light-impenetrable cloth in the field. Figures 6.1.3 and 6.1.4 show measured spectra under different cloths.

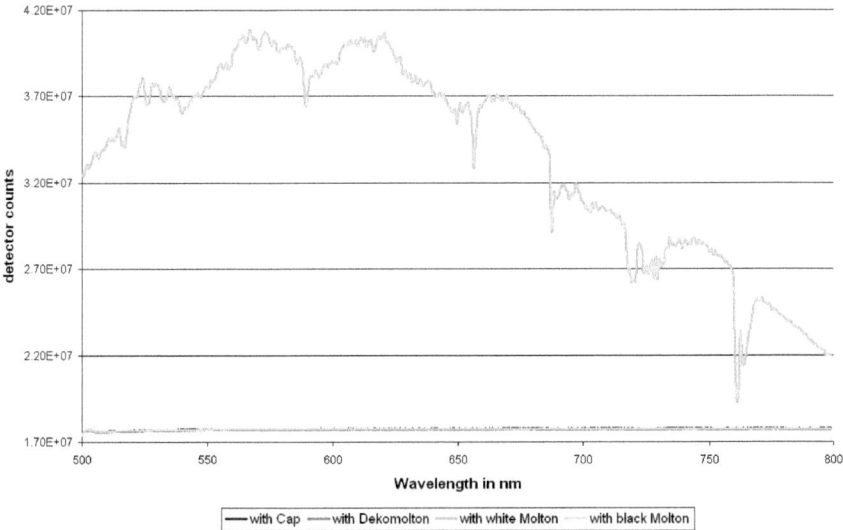

Figure 6.1.3:
Measured spectra when pointed at the midday-sun with the entrance covered by different cloths. It is immediately obvious that the white Molton (green line) is unsuitable for our purposes.

The test results showed that only one cloth, the Dekomolton, was effectively impenetrable to light.

6.1.3 Methods for creating reference spectra

All measurements of reflectance of vegetation included into our reference data set were produced using artificial light. The exclusion of natural light was ensured in different ways, depending on the specific requirements of the measurement.

Tent
For many of the grasses we measured directly in the botanical garden without disturbing the growing plant. For those measurements we erected a little tent, built from the light-impenetrable-cloth draped over a tripod. The artificial light-source was hung from the tripod while the MiniMax-DOAS was positioned in nadir over different parts of the grass-plot to allow for averaging later. The measurement was started by a person outside the tent so that the light radiating from the computer-screen did not influence the results.
This method was limited to plants fitting underneath the tent i.e. those with a height of up to 1.5m, the maximum height of the tripod.

Figure 6.1.5: The measurement tent set up in the botanical garden of Mainz University from the outside (A) and from the inside (B).

Closet
For the biggest grasses (e.g. Zea mays) and trees we collected fresh leaves and brought them into the office where a closet was fitted for measurements under dark conditions. The MiniMax-DOAS was in nadir and the artificial light source was fixed in near nadir position while the leaves were placed on black velvet of especially low reflectivity. The computer was again placed outside the dark area.

Blacked-out office
For measurements that required more room (e.g. measuring the effect of changing angles of instrument and light source), we blacked out the complete office. For those cases we used

thick, light impenetrable cardboard to cover and effectively black out the windows. The background for the measurements was again covered by the velvet of especially low reflectivity. All light sources were switched off except the screen of the laptop needed for the measurements. The screen was positioned as far as possible from the measurement location and pointing into the opposite direction.

6.2 Measurements over Vegetation

Vegetation is highly variable, even within the same species or even between different leaves from the same plant, not to mention variations during the growing season. Thus one single measurements would be far from representative. To limit the number of measurements to a manageable number, the general measurements were done over mature grasses and leaves/needles in good condition. Only some special measurements included other specifications.

To allow averaging for a species, measurements were done over different parts of the grass-plot or different parts of leaves/needles (close to the tip, close to the bottom etc.). If possible, both leaves with high and low exposure to light were collected.

At the start of each session, the offset and dark current of the MiniMax-DOAS were measured.

6.2.1 Single Measurement

Light-reference
The characteristics of the artificial light-source have to be recorded by measuring the reflection of the light over a reflector that does not significantly absorb within the spectral range of interest. In this case we measure from 500 to 800 nm. Grainy salt produces the desired reflection.

Measuring Reflection over Vegetation
To measure the reflection over vegetation, light source and MiniMax-DOAS are positioned in nadir. The reflection of the artificial light source is recorded by the spectrometer.

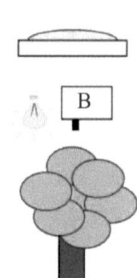

Due to the different height of the various grass-species, the distance from the light source to the measured vegetation is highly variable. Since the power of light is reduced by the square of the distance, the exposure time for each measurement has to be adjusted individually to generate a signal of about 80 to 90% of the maximum possible detector capacity. This allows for a signal strong enough to separate clearly from noise but not risk saturation of the detector.

The number of scans was always set to 100. The resulting averaging serves to keep background noise as low as possible without prolonging the measurement too much. (The Halogen lamp becomes hot over time; long exposure of vegetation in close proximity to high temperatures may cause damage to the plant.)

Figure 6.2.1: measuring reflection of artificial light-source over salt (A) and (B) measuring reflection over vegetation

Ratio
After correcting for offset and dark current the spectrum measured over vegetation is divided by the spectrum over salt. The resulting ratio shows the reflection properties of the studied plant.

Logarithm
The logarithm of the ratio is calculated because the Lamber-Beer-Law describes an exponential relation.

High Pass Filter
A high pass filter (FWHM: 7.8 nm) is applied next, eliminating the low frequency structures.

Smoothing
To smooth out noise, a low frequency filter (FWHM: 1.7 nm) is applied.

6.2.2 Averaging over a single Plant

As mentioned before, the spectral features of a plant may vary over different areas of the same plant. Therefore several spectra were taken over each plant or grass-plot, always measuring over different spots. Those single spectra were combined to produce an average over the plant (after correcting for offset, dark current and light-source). The resulting spectra were then further treated as described for single measurements.

6.2.3 Averaging over Plant-Groups

The averages over single plants were grouped into groups according to family or plant-group with similar spectral features. Those groups of similar spectral features were sometimes subgroups of families and sometimes grouping together families with very similar spectral features. This does not indicate a genetic or ecological connection.

6.2.4 Resolution Adjustment to Satellite instrument

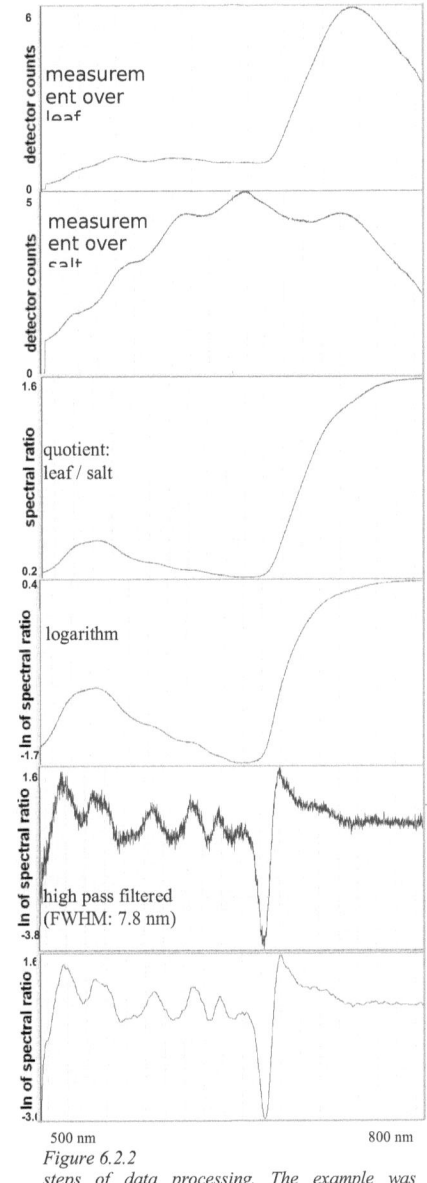

Figure 6.2.2
steps of data processing. The example was measured over a beech-leave.

The spectral resolution of the Mini-MAX-DOAS instrument is different from the satellite instruments. For each satellite-instrument, the reference spectra need to be adjusted to the individual spectral resolution and slit-function.

6.3 Vegetation Reference Spectra

6.3.1 Initial averages for first trials

For our initial investigations we only applied six different vegetation classes to establish the amount of separability we will be able to get. For more results see Chapter 9.

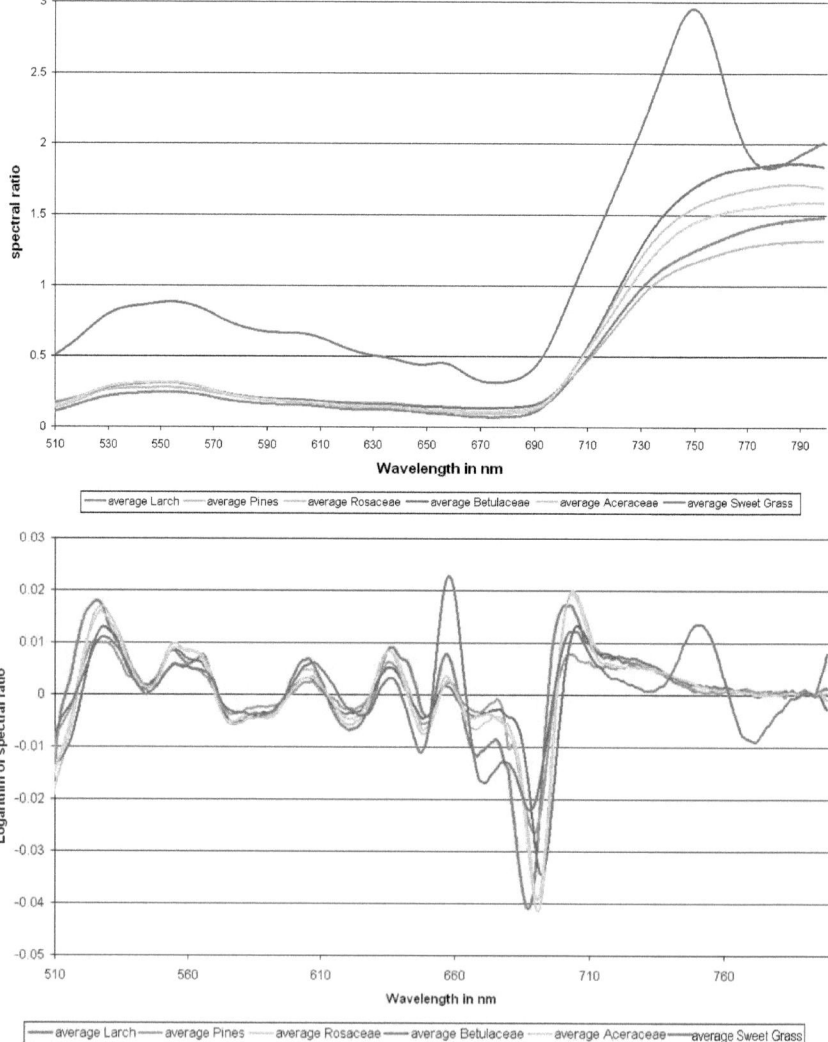

Figure 6.3.1
This graphic shows the original (top) and high-pass filtered (bottom) reflectance spectra of the initial vegetation-classes. Trees are still in single families. Larch and Pines are coniferous, Rosaceae, Betulaceae and Aceraceae are deciduous. Sweet Grass only contains millet and maize here.

Our initial six classes were Larch, Pines, Rosaceae, Betulaceae, Aceraceae and Sweet Grass. Initial results suggested combining classes of similar spectral characteristics, like e.g. Betulaceae (Birch family) and Rosaceae (Rose family, including many fruit trees).

To include more of nature's variability, many more plants were measured. Wherever measurements over several species of a family could be established, averages were produced. The spectral characteristics were investigated and groups of families were formed where applicable. Thus refined reference spectra were produced.

6.3.2 Refined reference spectra

The initial investigations made clear that some reference spectra are too similar in characteristics for separation into different classes. So bigger groups, according to reflectance characteristics, were produced. The content of these groups is presented in Figure 6.3.2. The spectral characteristics are shown in different graphics, first presenting deciduous, then conifer and finally grass signatures.

Content of the new reference spectra:

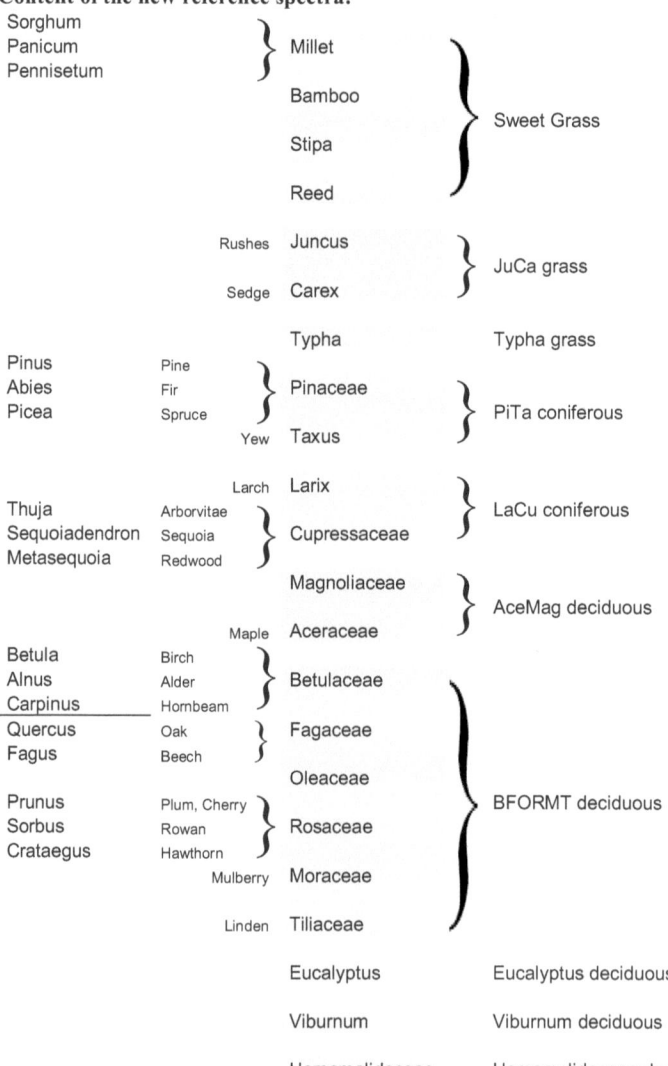

Table 6.1: Content of the refined spectral reflectance vegetation-classes.

Deciduous trees have very different kinds of leaf buildup. (Compare Chapter 2.) Spectral characteristics vary accordingly. (See also e.g. Chapter 8.2.) Figure 6.3.2 presents the new reference classes for deciduous vegetation. Some families have very specific spectral reflectance characteristics (e.g. Viburnum at 660-680 nm), others share similar signatures with many other classes, like those grouped together as BFORMT. The families forming this class are Betulaceae, Fagaceae, Oleaceae, Rosaceae, Muraceae and Tiliaceae. Compare also Table 6.1.

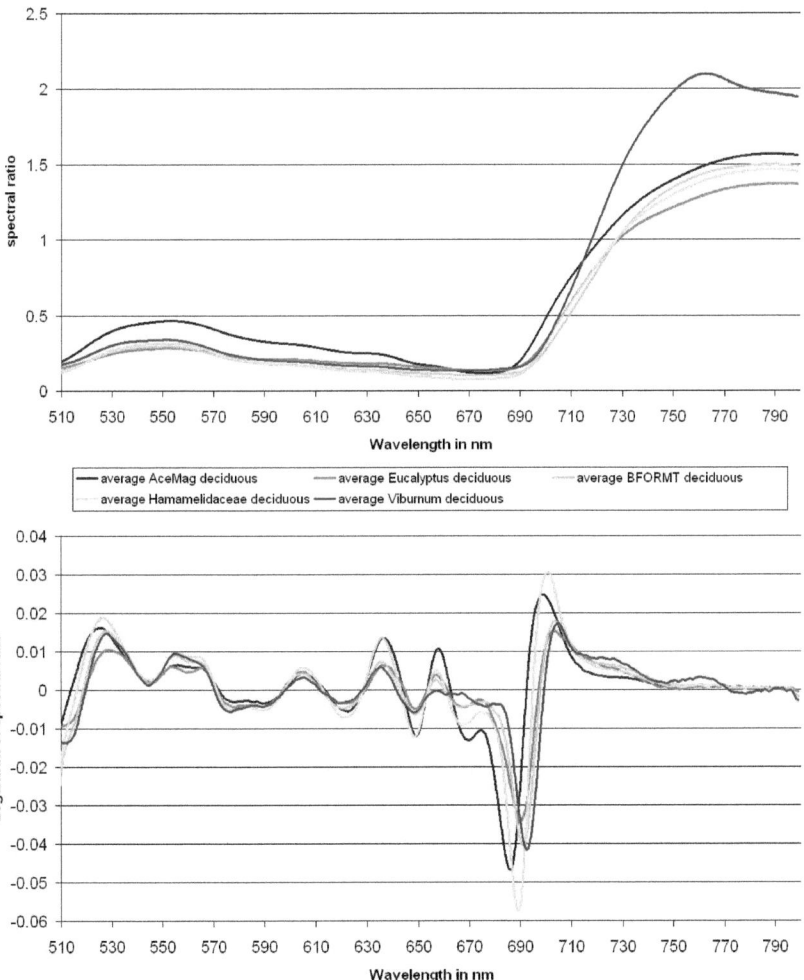

Figure 6.3.2
Original (top) and high-pass filtered (bottom) reflectance spectra of the refined deciduous classes. AceMag and BFORMT are classes combining deciduous families with very similar reflectance characteristics. The actual content of the different classes is given in Table 6.1.

The new reference classes for conifers do not strictly follow plant-families. Pinaceae combine many groups of far-spread conifers (spruce, fir, larch, pine). Thus the spectral signature for the whole family becomes too diverse for averaging. Larch had to be treated separately. It shows a similar spectral response as Cupressaceae, with which it is combined as class LaCu.

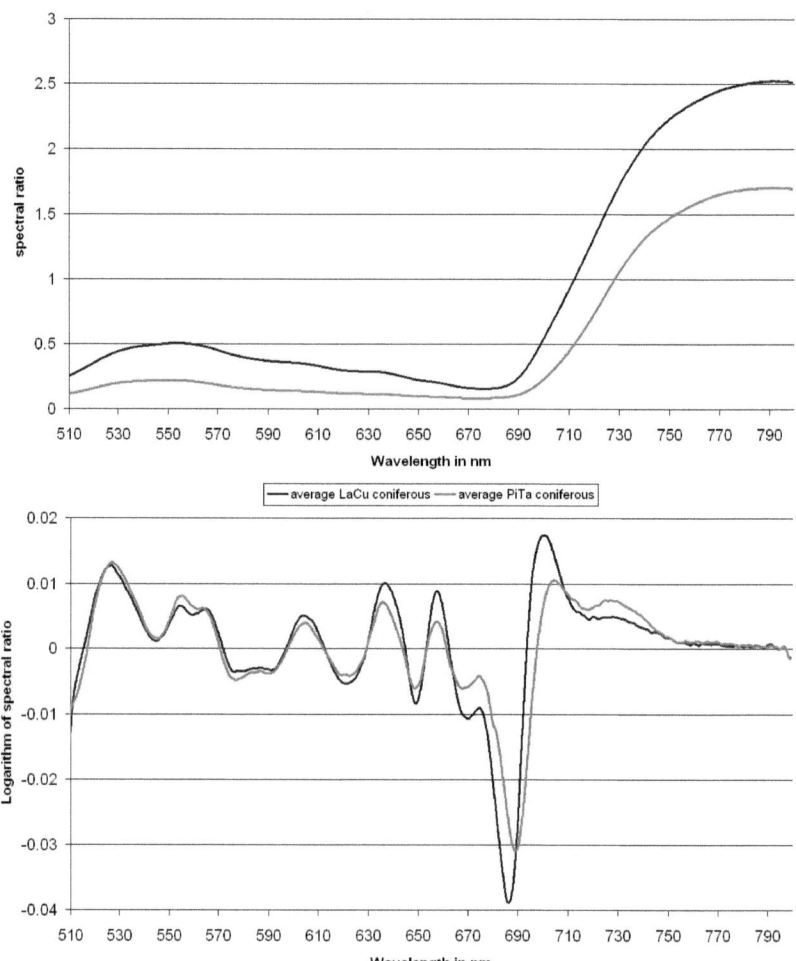

Figure 6.3.3
Original (top) and high-pass filtered (bottom) reflectance spectra of the refined coniferous classes. Both classes are combinations of different families. The actual content of the different classes is given in Table 6.1. Pinaceae are in the class PiTa, except for Larix, whose spectral characteristics better combine with

Sweet grasses are a very large group and contain all the grasses important for food production.

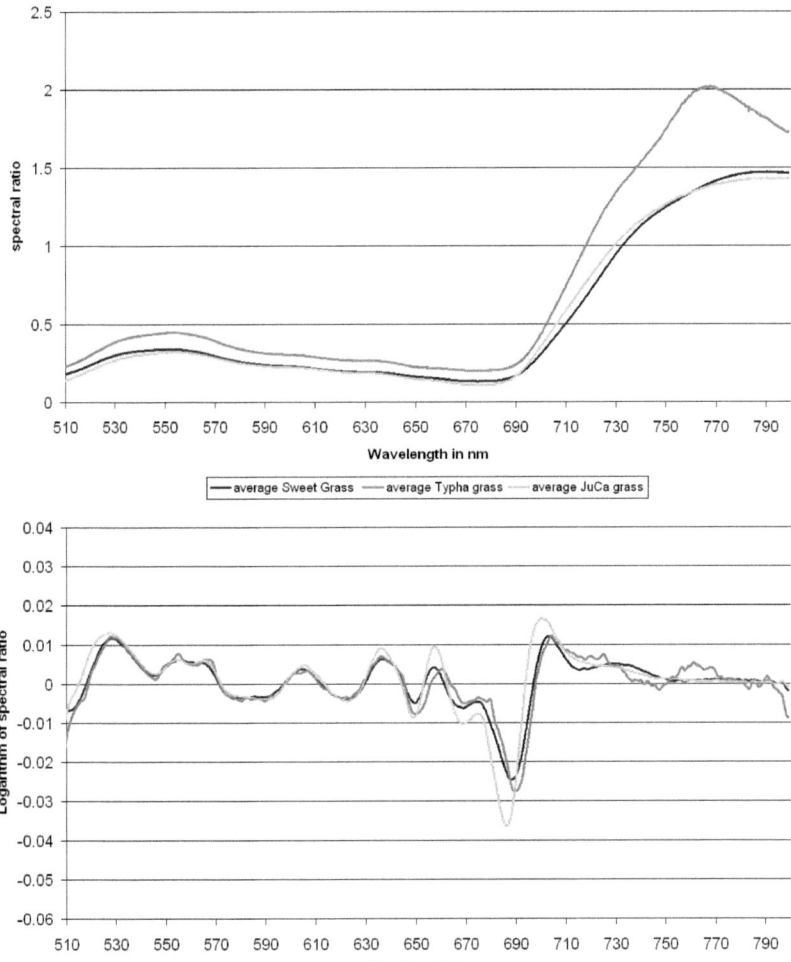

Figure 6.3.4
Original (top) and high-pass filtered (bottom) reflectance spectra of the refined grass classes. The actual content of the different classes is given in Table 6.1. Sweetgrasses measured for this thesis all responded so similarly that they all occupy the same class, no matter to which climate-zone they are adapted. Both other classes mostly prefer wet environments.

According to the measurements presented here, their spectral signatures unfortunately do not offer themselves for separation. Typha show a very specific signature, justifying their own class. Sedge and rushes form a common group: JuCa. JuCa and Typha contain species that mostly favour moist to wet growing conditions.

7 Initial analysis of satellite spectra using GOME

Before we started to produce our reference data, we had to get a general overview of the ability of DOAS to measure vegetation signals. For those first trials vegetation spectra form the ASTER-spectral library were used. The vegetation classes used were conifer, deciduous and green grass. These were applied to GOME-data.

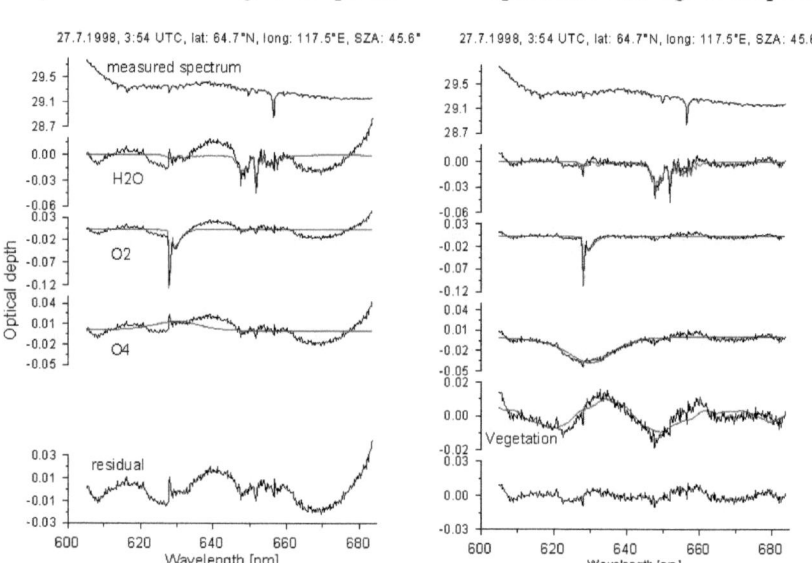

Figure 7.1
Comparison of DOAS fit without and with vegetation spectra. The residual of the fit without vegetation-spectra shows clear spectral features that suggest an additional important absorber. The residual after including vegetation is clearly more random. From T. Wagner et al.: Atmos. Chem. Phys., 7, 69-79, (2007)

Wagner et al. (2007) first interrogated into the potential of the red spectral range for vegetation monitoring with DOAS, when they took a closer look at the residual of the H_2O analysis. Vegetation especially appeared to have an impact, as demonstrated in Figure 7.1. When only fitting water vapour, O_2 and O_4, to the measured spectra, the residual showed structures over land with high vegetation content. To check the theory that this might be a vegetation signal, they fitted also vegetation spectra analogue to a cross-section. (See also Figure 3.2.3 in Chapter 3.) The applied vegetation spectra were from the ASTER spectral library. The resulting residual was much smaller.

7.1 Clouds

The influence of clouds on satellite data is well known. In the optical range clouds basically hide the lower atmosphere and the ground from the satellite instrument. Since we are interested in vegetation which is growing on the ground, this shielding effect has to be taken into account. As the pixel size of the satellite instruments used covers many kilometres, a certain fraction of cloud-cover is almost always inevitable. To investigate the influence of cloud fraction on our results, we calculated monthly means for grass and deciduous spectra from the ASTER-spectral library on GOME-data. Figure 7.1.1 shows the results for the months June to August 1996.

When assessing the monthly means only pixels of up to 10% and 20% cloud-cover were compared with monthly means including all pixels without consideration of cloud-cover. Missing data appears white in the pictures.

Filtering out pixels with high cloud-cover strengthens the vegetation readout considerably in the monthly averages. We show this in Figure 7.1.1 for deciduous and grass spectra for the months June to August, when vegetation in the northern hemisphere grows strongest. This strengthening effect is visible e.g. in the deciduous spectra in June in Figure 7.1.1. Compared to the monthly-means with all measurements, the image with max. 10% cloud-cover displays much higher signals in Eurasia, the Americas and Africa south of the Sahara. The effect is even more pronounced for grass spectra.

The strength of the vegetation response is strongest in the pictures including only pixels with lowest cloud fraction, but the number of missing pixels is highest here. This shows the effect of the pixels with high amounts of shielding, which – when included in the mean – considerably reduce the vegetation signal. Since the number of missing pixels is rather high when only including pixels of maximum 10% cloud-cover, for general mapping purposes a maximum cloud-cover of 20% appears to be a good compromise.

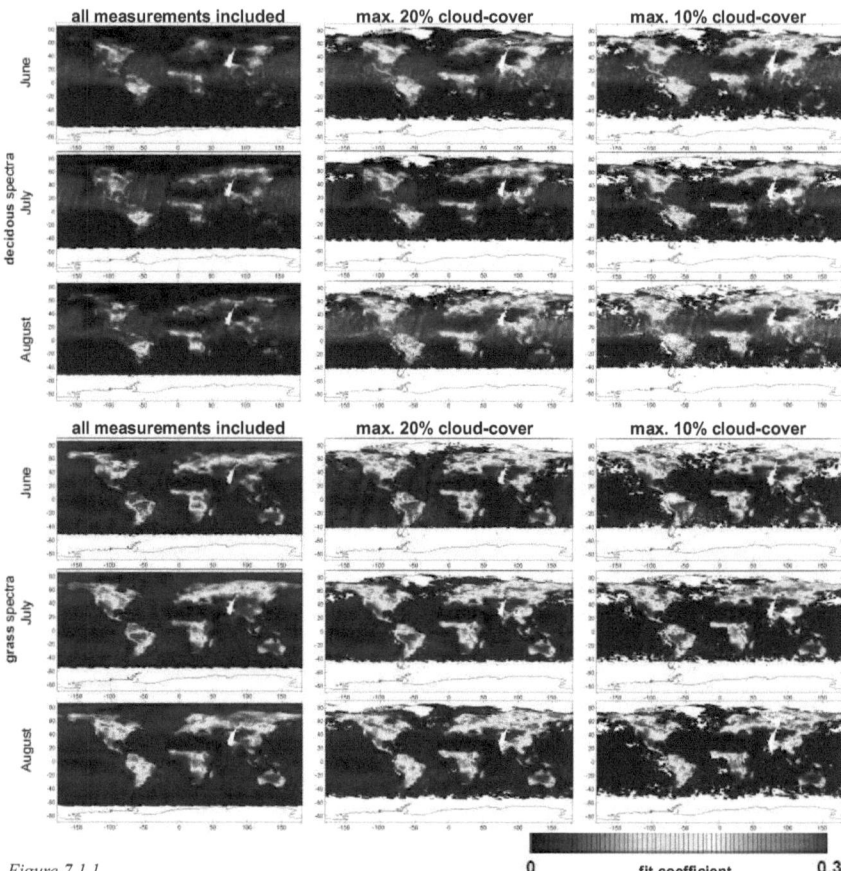

Figure 7.1.1
Using HICRU to account for cloud-effect. Monthly means of the retrieved fitting coefficients from GOME data for fits using the logarithm of the vegetation spectra from ASTER Spectral Library. The monthly means on the left side include all pixels. The resulting vegetation-signal is considerably lower than in the monthly means of the images choosing only pixels with lower cloud-cover. The lower the cloud-cover, the higher the number of missing pixels (white areas), showing that no data was recorded within that month with the required low cloud-cover.

The shielding effect of the pixels with higher cloud-cover is demonstrated in more detail in Figures 7.1.2 and 7.1.3.

Comparison all pixels against cloud-cover 10% for conifer (10/97)

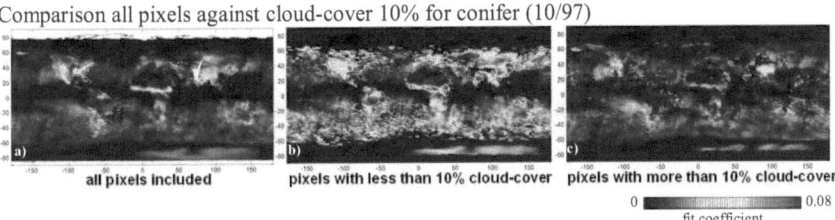

Figure 7.1.2
Comparison between monthly-mean (fitting coefficients for the logarithm of the conifer spectra) images using a) all pixels, b) pixels with max. 10% cloud-cover and c) more then 10% cloud-cover. The conifer-signal is considerably stronger in b) than in a). C) still shows some conifer-signals penetrating cloud-cover but also showing large areas of no signal where conifer-signal is strong in b). This shows how clouds are masking out vegetation-signal.

Comparison all pixels against cloud-cover 20% for conifer (10/97)

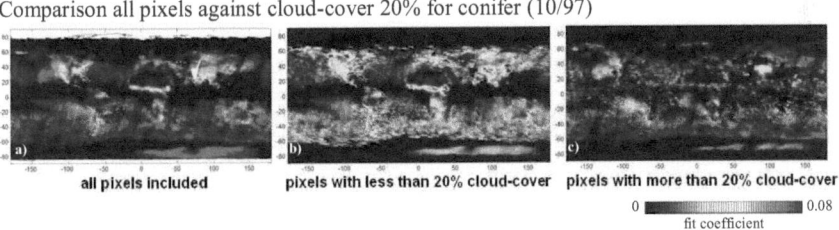

Figure 7.1.3
Comparison between monthly-mean (fitting coefficients for the logarithm of the conifer spectra) images using a) all pixels, b) pixels with max. 20% cloud-cover and c) more then 20% cloud-cover. The effects are similar to those shown in Figure 7.3.2 but this time b) includes fewer missing pixels. This more complete image is paid for with a clearly reduced conifer-signal. Compare e.g. the signal for Africa south of the Sahara.

7.2 Bias correction for vegetation analysis

Like in many DOAS retrievals for satellite observations, also temporal varying biases were found for the GOME vegetation analyses. Thus the data must be calibrated to 0, in our case no vegetation. For this purpose areas which can be considered vegetation-free have to be used. These calibration areas have to be sufficiently large to include several ground pixels. For this thesis I investigated an area over the Sahara Desert and over the southern Atlantic Ocean.

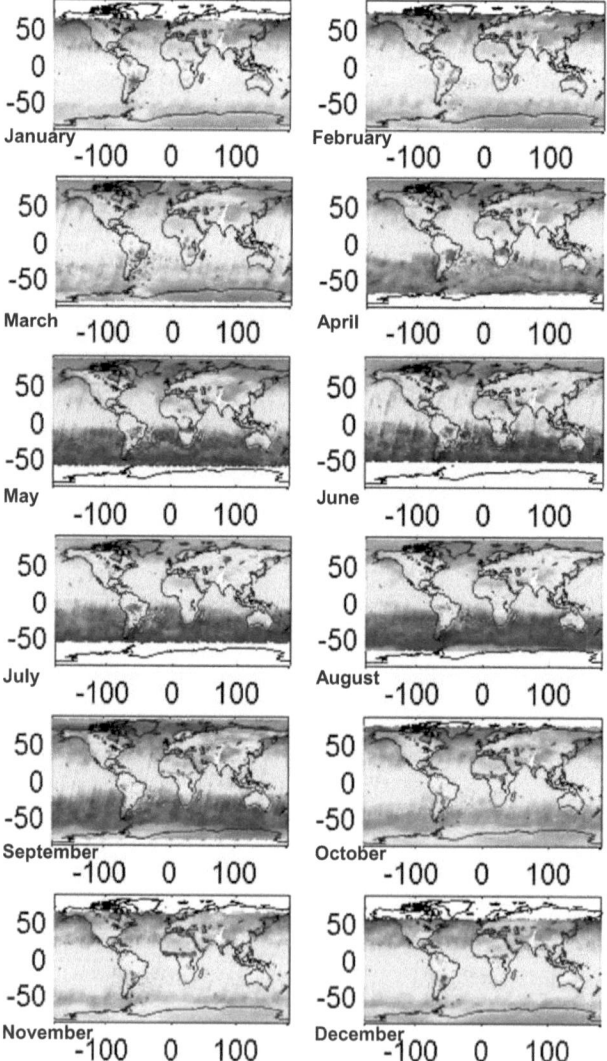

7.2.1 Sahara

The Sahara Desert can be considered as an area without vegetation. The spatial dimensions of the oases should be sufficiently small, so that the resulting vegetation signal can be ignored.

We defined the reference area as being between 18°-30°N and 5°W-30°E. Figure 4 demonstrates the results for deciduous vegetation after applying the bias correction in monthly means for the year 1998.

Figure 7.2.1
Sahara set as zero vegetation reference for monthly averages 1998 for deciduous-signal, no cloud-correction.

It is apparent that the calibration works well over known desert areas. However, it does not correct for apparent vegetation signals in the tropical oceans.

7.2.2 Southern ocean

The southern ocean can be considered free of vegetation and mostly free of algae. The occasional algae bloom should be small enough not to interfere too much. The reference-area was chosen to be all the water between 56 and 60° south. Figure 5 shows monthly means for the deciduous signal for 1998 after applying this bias correction.

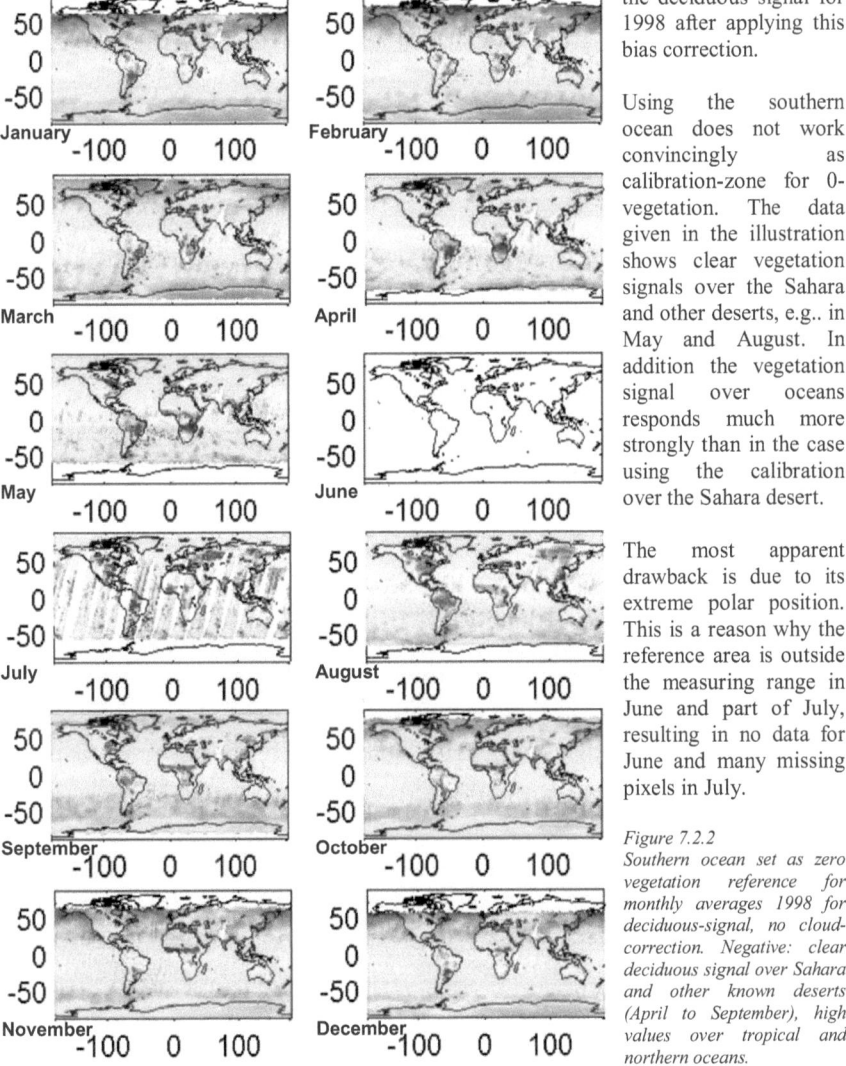

Using the southern ocean does not work convincingly as calibration-zone for 0-vegetation. The data given in the illustration shows clear vegetation signals over the Sahara and other deserts, e.g.. in May and August. In addition the vegetation signal over oceans responds much more strongly than in the case using the calibration over the Sahara desert.

The most apparent drawback is due to its extreme polar position. This is a reason why the reference area is outside the measuring range in June and part of July, resulting in no data for June and many missing pixels in July.

Figure 7.2.2
Southern ocean set as zero vegetation reference for monthly averages 1998 for deciduous-signal, no cloud-correction. Negative: clear deciduous signal over Sahara and other known deserts (April to September), high values over tropical and northern oceans.

The reasons for the exclusion of areas with high solar zenith angle are discussed in Chapter 3. From here on the bias correction is done against the area in the Sahara Desert.

7.3 Summary and Conclusion

Initial investigations show the potential of DOAS to study global vegetation. However, the results so far are not satisfactory. Compare e.g. the high grass-response in June and July in the Amazon-Basin (Figure 7.1.1).

This may be caused by the used reference spectra. The ASTER Spectral Library was not collected for DOAS-applications, serving satellites with lower spectral resolution. In addition the limitation to only three vegetation classes might be too simple.

Therefore we have to produce our own vegetation reference spectra.

Since DOAS analyses the high-frequency structures of reflectance, we also have to investigate potential influences on these structures under different conditions.

8 Results of the Mini-MAX-DOAS measurements

Detailed observations of vegetation reflectance were performed using the Mini-Max-DOAS instrument, including different characteristics of leaves, like different colours of leaves, the front- and backsides, but also the influence of changing angles of the light source and the measurement instrument. In this Chapter we first start off with influences on leaves and then continue on to the effects of instrument and light angles.

The investigated material was collected on the Campus, mostly inside the Botanical Garden, of Mainz University. We measured either on site or inside the office, which is situated right beside the Botanical Garden. Only small numbers of samples were taken to be measured immediately. This was repeated, if necessary, several times per day. Thus, the material was always fresh. Measured plants are documented in Appendix A.

If not stated otherwise, we measured the reflectance of the upper side of the leaves, since these are predominantly directed towards satellites. The Mini-MAX-DOAS was in nadir, the artificial light source as close to nadir as possible.

8.1 Leaf with and without pigments

The first study investigates the importance of pigments and leaf structure for vegetation reflectance in the 500 to 800 nm wavelength range. To see the different influences of structure and pigments of the leaf on reflectance characteristics, a plant with white and green leaves was used. For this we needed a plant with spotted, two-coloured leaves. Those usually also produce one-coloured leaves occasionally. (No plant could afford too many white leaves without risking starvation.) The plant found in the botanical garden having these characteristics was Hedera colchica, a species of ivy well-liked as an ornamental in many gardens.

We measured a white and a green leaf under the same measurement conditions. Figures 8.1.1 and 8.1.2 show the results. The blue line denotes the reflectance of the white leaf and the green line the reflectance of the green leaf. The green leaf shows the typical broadleaf characteristics, but the white leaf shows some significantly different structures which are obviously masked when dealing with a coloured leaf.

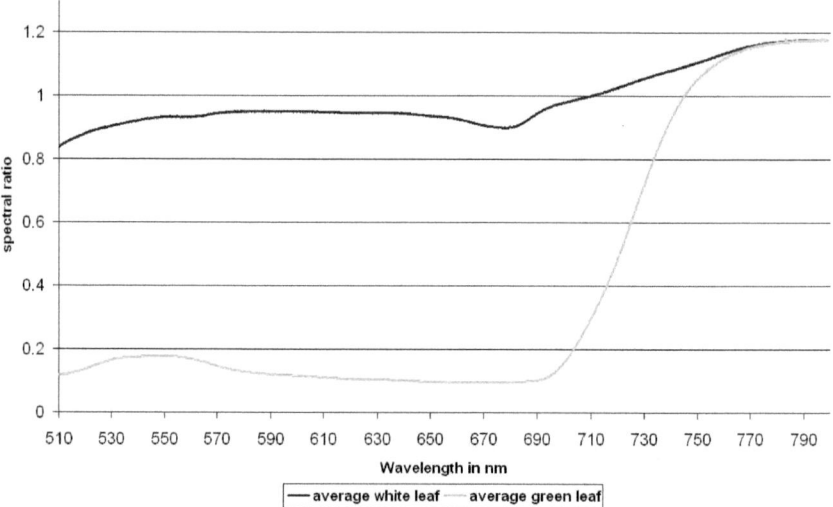

Figure 8.1.1
Reflectance of white and green leaf of Hedera colchica. Reflectance of the white leaf in the visible light is generally much higher than of the green leaf. Some light absorption is visible, interestingly in different positions compared to the green leaves. From 770 nm onwards the reflectance of both leaves is the same.

In Figure 8.1.1 the strong reflectance of the white leaf in all wavelength-ranges is apparent. Only at 770 nm and above do reflectances of both leaves converge. A slightly reduced reflection is visible at 550 to 565 nm, another at about 670 to 685 nm. A stronger reduction begins to show towards the blue spectral range. The general signal confirms the findings presented in Figure 2.2.3 (Chapter 2).

Especially interesting in Figure 8.1.2 appears to be the absorption around 560 nm and another absorption structure at around 680 nm. Since the leaf was absolutely white, these absorption lines must be caused by either the structure of the leaf or by internal cell organelles. Further investigation into this topic might be interesting and reveal the real sources of the features of the white leaf.

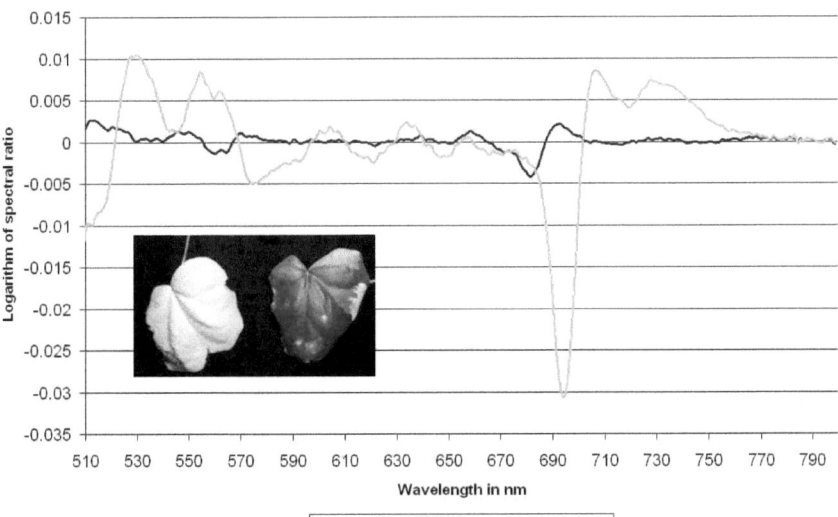

Figure 8.1.2
High-frequency reflectance structures of white and green leaf of Hedera colchica The amplitude is very much reduced in the white leaf. Apparent absorption-structures are in different positions compared to the green leaf.

8.2 Reflection from upper and lower side of a leaf

Leaves have a significantly different build up on the upper side and the lower side. Some leaves show this more than others (see also Chapter 2.2.2.2). To see how this influences scattering and absorption characteristics, the reflectance over the upper- and the lower side of six different leaves was measured. These leaves were chosen from different plant-families and displayed different characteristics. The characteristics range from the hairless and relerlatively flat upper and lower side of Prunus to the structured Viburnum with some hairs on the top and lots of hairs on the lower side. The average over these leaves for both sides was calculated.

It is immediately apparent that the characteristics of upper- and lower sides are very different. Also the colours to the eye differ much. All upper sides are darker compared to the lower sides, as a result first of the higher chlorophyll-content in the cells closest to the epidermis and second of the intensified light-scattering caused by hairs or other structures on leaves with apparent whitish colours on the underside. This is shown in the measurements. First the average upper- and lower sides are considered. The average lower side reflectance is much reduced in comparison to the upper side in all wavelength ranges. All wavelength ranges are significantly reduced in signal. Interestingly, for high-pass filtered data from about 770 nm onward, all reflectances converge.

Figure 8.2 also shows the single measurements in the background. These show the variability of the signals.

The different behaviour of upper- and lower side reflectance is important to be considered for satellite measurements, because the satellite does not always see only the upper side of deciduous leaves. As shown earlier in Chapter 2.2.2.2, plants may actually actively change the aspect of the leaf towards the sun, exposing the lower side for protection against too much incoming sunlight. Another factor might be wind that occasionally turns the leaves in a tree upside down. A satellite overpassing during that time would register a considerable amount of reflectance from the lower side of the leaves.

Since this thesis presents only first results of applications of the new vegetation reference data for DOAS satellite retrievals, the investigations were restricted to reflectance-signals of the upper sides of leaves. However, this topic should be further investigated in future.

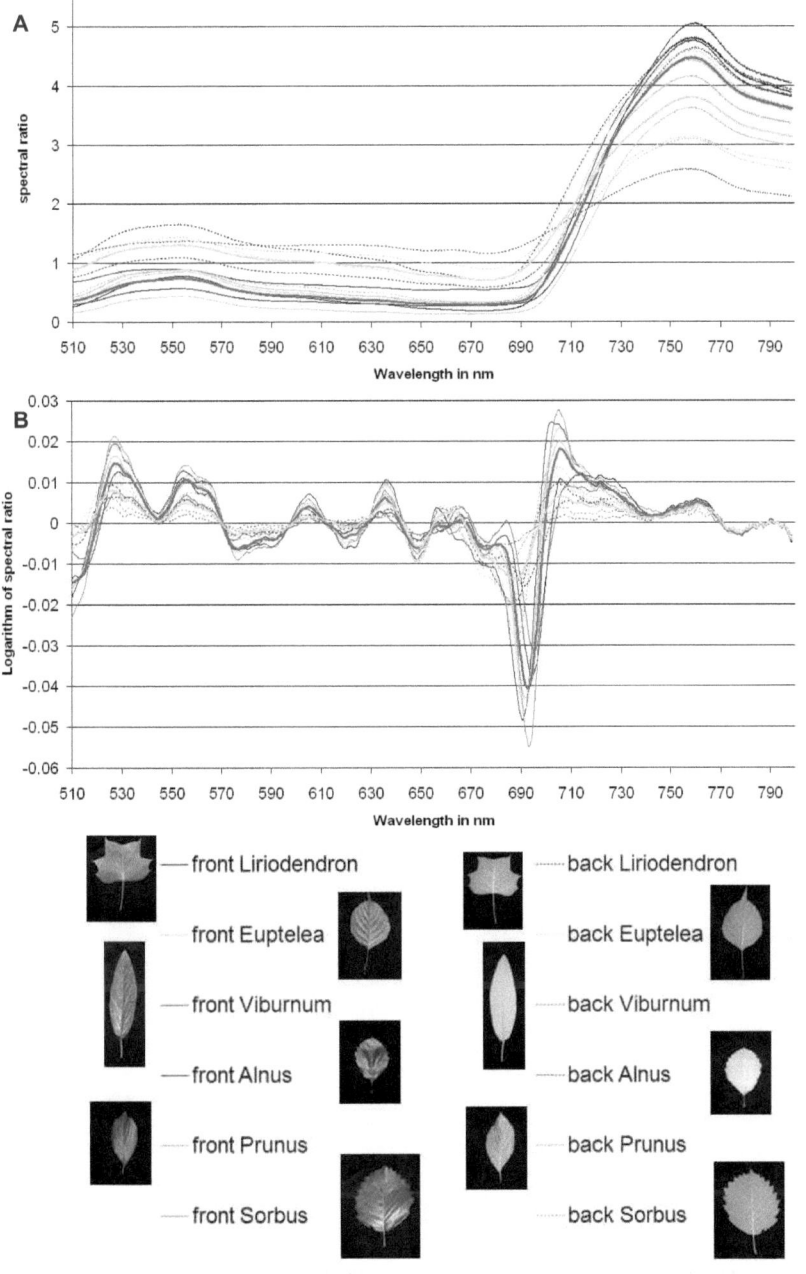

Figure 8.2.1
Comparison of reflection from front and back of leaves. Leaf fronts have different characteristics from flat and matt (Prunus), flat and glossy (Alnus) to structured and slightly hairy (Viburnum). Backsides range from hair-free (Prunus) to very hairy (Viburnum). Averages show a significantly reduced absorption on the backs. Average spectra converge only at about 750 nm. Single measurements in the background show the range of variability between different kinds of leaves. A shows the spectral ratio, B shows the high-frequency structures on the logarithm of the spectral ratio.

8.3 Comparison of spruce-needles

Coniferous trees, with the exception of the larch family, keep their needles for more than just one year. In this Chapter two characteristics of spruce are investigated. The first question was if there is a difference between needles grown in the year of measurements and needles from the year before, still attached to a branch. The second question was if there is a difference in the reflection signal if the needles are attached to a branch or if they are separated from the branch. For this study we used samples from a spruce tree, since this species is rather common on campus in Mainz and could also be sampled outside the botanical garden. To remove big samples from species inside the Botanical Garden was out of the question.

Measurements over needles attached to the branch were done first, immediately followed by removing the needles and measuring the detached samples. Thus, differences in reflectance should only result from changes in measurement-geometry.

Figure 8.3 shows the two measurements over needles on a branch in dark-blue and pink. The results for the needles removed from the branch are displayed in green and light-blue. The first significant difference in the high-frequency data is the reflectance at about 540 nm, where the detached needles have the same absorption band as broadleaf leaves while the needles attached to the branch have a different signal. This counts for both ages of needles. This might actually be caused by the branch that is visible through the needles and obviously has some influence on the reflectance signal. Another significant difference between attached and non-attached needles is that the non-attached needles show a clear reflection structure near 680 nm, whereas the attached needles only show a shoulder there.

A surprising difference between this year's attached and detached needles is that the absorption of chlorophyll a (about 680 - 700 nm) appears to be the same, whereas the needles from the year before show a significant difference in the absorption of chlorophyll a.

The general difference in reflectance of older and younger needles attached to the branch is also visible, especially beyond 630 nm, where the absorption structures and the reflectance structures of the older needles are much more pronounced. Here the observation is simply reported. We do not have explanations yet. However, further research into this question could be interesting.

Why is this important for satellite measurements? The way coniferous trees are structured, the parts of the branches most exposed to satellite registration are the youngest because they are those furthest on the outside. However, a significant amount of the signal also comes from the needles less exposed but still visible from the top. Thus, the signal from a spruce-tree will always be a mixture of the signals from the older and the younger needles. According to these measurements a true average over a coniferous tree has to include needles of different ages. The differences between attached and non-attached needles also suggests that a true signal for satellite retrievals should be produced from needles which are still attached to the branch to include all the characteristics.

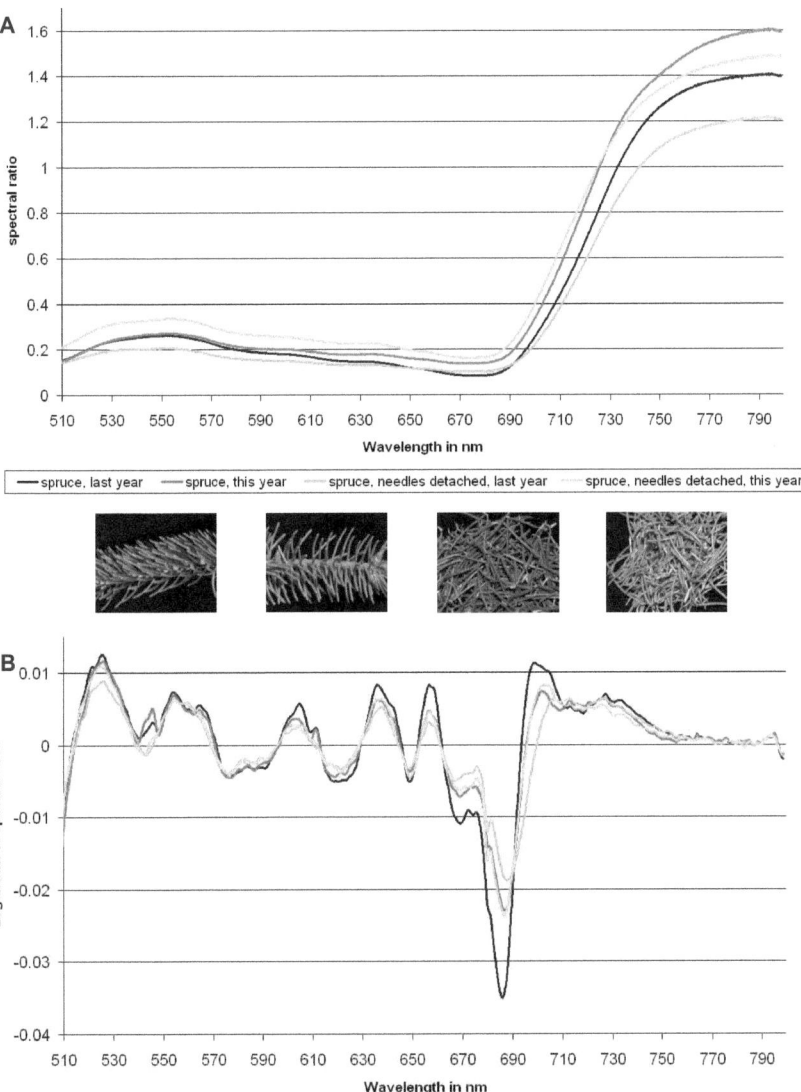

Figure 8.3
Comparison of spruce needles. Needles were investigated attached and non-attached and at two different ages (this year and last year). A shows the spectral ratio, B the high-frequencystructures of the logarithm of the spetral ratio.
Needles attached to the branch show a difference from non-attached needles at about 540 nm. Otherwise this year's needles don't seem to show significant differences between attached and non-attached. This is different with last years needles, whose spectra diverge from about 600 nm onwards, especially apparent in the major chlorophyll absorption between 660 and 700 nm.

One explanation for the difference in the signals of attached and non-attached needles might be that non-attached needles all lie flat, giving a similar angle of incident light and measurement instrument to all needles. Needles attached to a branch grow in clusters and therefore have slightly different angles towards instrument and light, producing different effects. Moreover, the branch connecting the needles is visible and is excluded if the needles are measured alone.

As a consequence, coniferous reference-spectra were measured with needles still attached to the branch, if possible. An exception was made for trees of the Pinus family. Their needles are so long and spread so far apart that the measurable area in the view of the instrument was too small. For measurements of Pinus the needles were detached and placed close together.

8.4 Changes over grass during vegetation-cycle

To be able to monitor the vegetation cycle over a grass plant, several plants of Panicum miliaceum - a common tropical cereal, whose seeds were kindly supplied by S. Alfonso of the Botanical Institute at Goethe University, Frankfurt a.M. - were grown under controlled conditions and measured almost weekly during the whole vegetation cycle. Measurements were done over several leaves of different plants. During measurement the leaf was positioned as close to horizontal as possible under the Mini-Max-DOAS in nadir. Due to the different sizes of the growing plant-leaves, the field of view was filled in varying percentages with plant and background (very dark soil).

The first measurement was done a week after seeding, when the first green parts of the plant were visible above the ground. It is quite apparent that the reflectance and absorption change significantly over time, not so much in the position of the absorption, but the strength of it. This is quite understandable, since the plant first starts off very small and has to develop its full capacity, growing tall and strong, producing more leaf area and better developed organelles for productivity. Then, following its growth pattern and cycle, it matures and reduces photosynthetic capabilities. During senescence it reduces photosynthesis in favour of producing seeds, allowing the vegetative part to dry out and die off. Like most grasses Panicum miliaceum has a one year-vegetative cycle and survives winter only as seeds. The reduction of photosynthesis is clearly visible in the loss of pigments, suggesting that the plant is actively breaking them down.

What does this mean for the measured signal? In Figure 8.4.2 this is most visible in the absorption of chlorophyll a (about 680 to 700 nm). The strongest absorption happens in weeks 7 and 8 (circle 1), where the plant is growing strongly. Circle 2 surrounds measurements of weeks 1 to 6, where the plant begins growing and starts developing. Then the slowly reducing levels (circle 3) show successively increasing age.

Conversely e.g. at about 520 to 540 nm the reflectance was again strong for the first five measurements and most extreme during weeks 7 and 8. Then again there is a slow reduction of the signal. The last measurements show almost no features, which is consistent with the situation of the plant, which is almost totally dry and dead at that point. (Compare photographs in Figure 8.5 with corresponding measurements.)

What does this mean for the DOAS-retrievals? The position of the absorption remains stable for most features except the main absorption of chlorophyll a, which moves several nanometres during the development of the plant. This is shown in Figure 8.4.1, comparing the signal of Panicum miliaceum of 7 and 9 weeks of age.

The strength of the absorption of the different wavelength ranges changes considerably. This has to be taken into account when interpreting measurements of grasslands from satellites. The same amount of grass may give a strong signal during its major growth-phase and a rather low signal during advanced senescence.

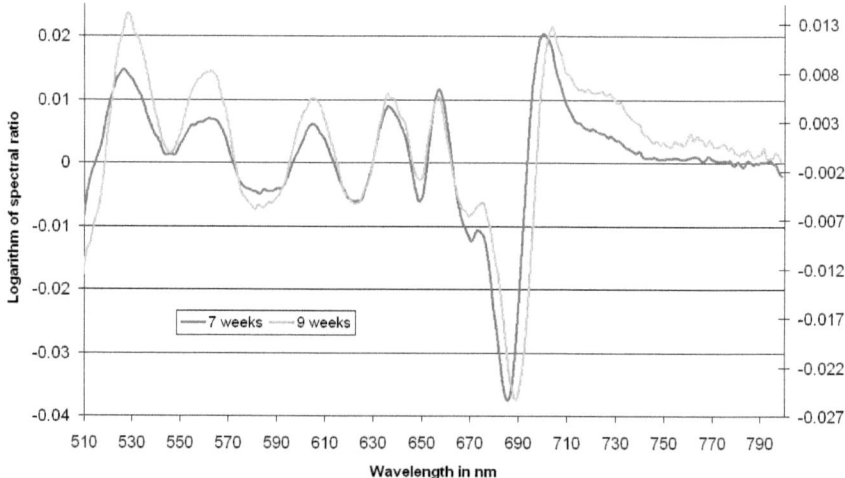

Figure 8.4.1
Comparison of reflectance of Panicum miliaceum of 7 and 9 weeks of age. Y-axis are adjusted to show similar amplitudes. Position of maximum absorption stays the same at all wavelengths except for the strongest absorption-peak of chlorophyll a between 680 and 690 nm.

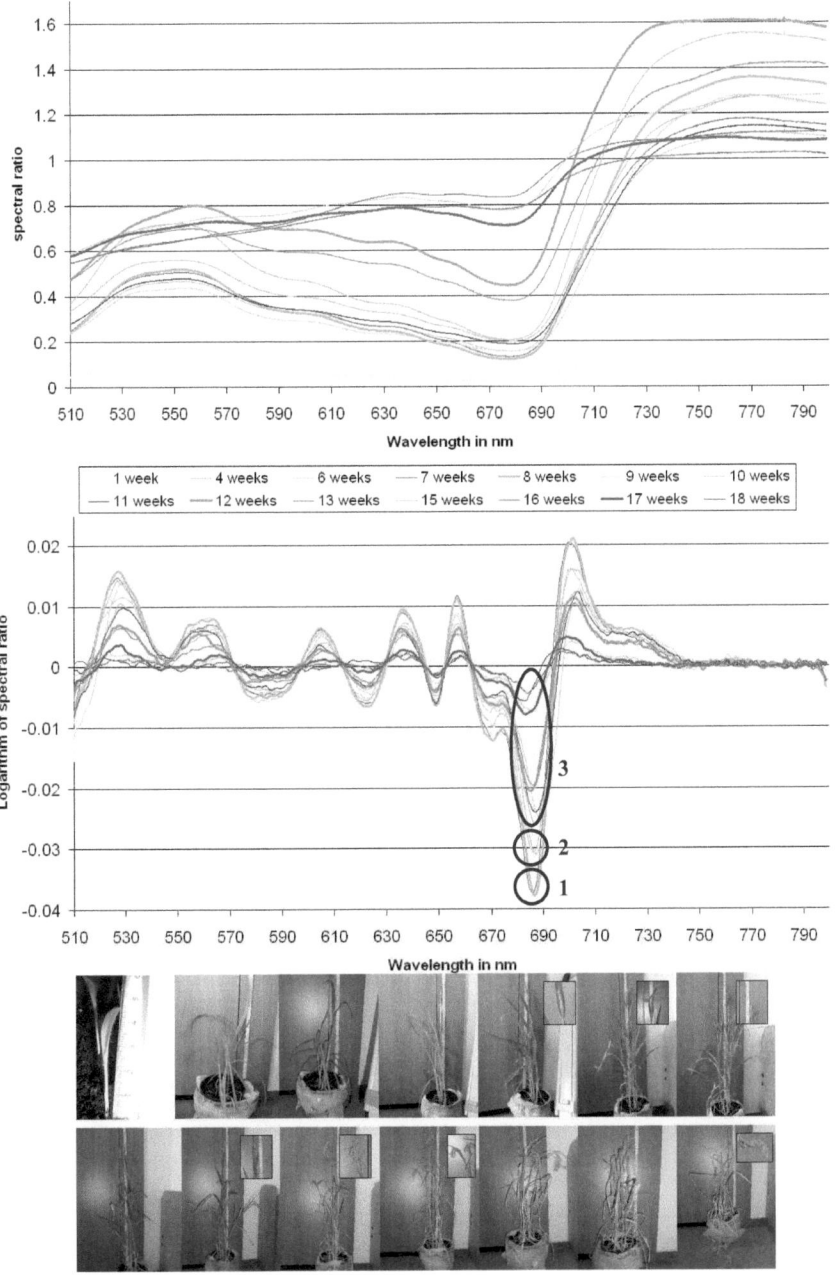

Figure 8.4.2
Several plants of Panicum miliaceum were monitored weekly during the vegetation-cycle. Missing weeks are due to technical problems. The legend shows weeks after seeding. For better readability every 4 measurements are shown with a wider line, progressing from yellow through orange to brown. The start of the vegetation-cycle shows intensifying absorption. Weeks 9 to 12 show slightly reduced absorption while weeks 13 to 18 show drastic reduction of the signal.

8.5 Changes in pigments and absorption in autumn

In autumn trees retrieve the material used for chlorophyll production. This process is responsible for the spectacular colours forests display at the end of the vegetation cycle. Some trees break down only the chlorophyll, changing their leaves from green to yellow before letting them go. Others, like the maple in this example, produce new pigments for a short time, turning the leaves orange and red as a result (compare shift of highest reflectance to longer wavelength in Figure 8.5 A). It is also clearly visible how strongly chlorophyll dominates the spectra. Wherever the green colour is still visible, the reflectance shows a pronounced relative low near the absorption peak of chlorophyll a. A little less obvious is the absorption peak of chlorophyll b near 650 nm.

Corresponding to the reflectance, the absorption visible in the highpass-filtered data in Figure 8.4.B shows a reducing chlorophyll absorption, best visible in the main chlorophyll a absorption around 690 nm, but affecting the whole range from about 590 to 700 nm.

The build-up of new pigments is seen best in the range of 550 to 570 nm, where dominating reflection in green leaves turns to absorption in red leaves.

The drying leaf (Maple leaf 10) clearly shows decreased infrared reflectance and a slight blue-shift of the red-edge, confirming the findings introduced in Figure 2.2.18.

The lines in the graph correspond to the leaves below.

The changing reflectance signals of autumn were measured for many trees (see Appendix A).

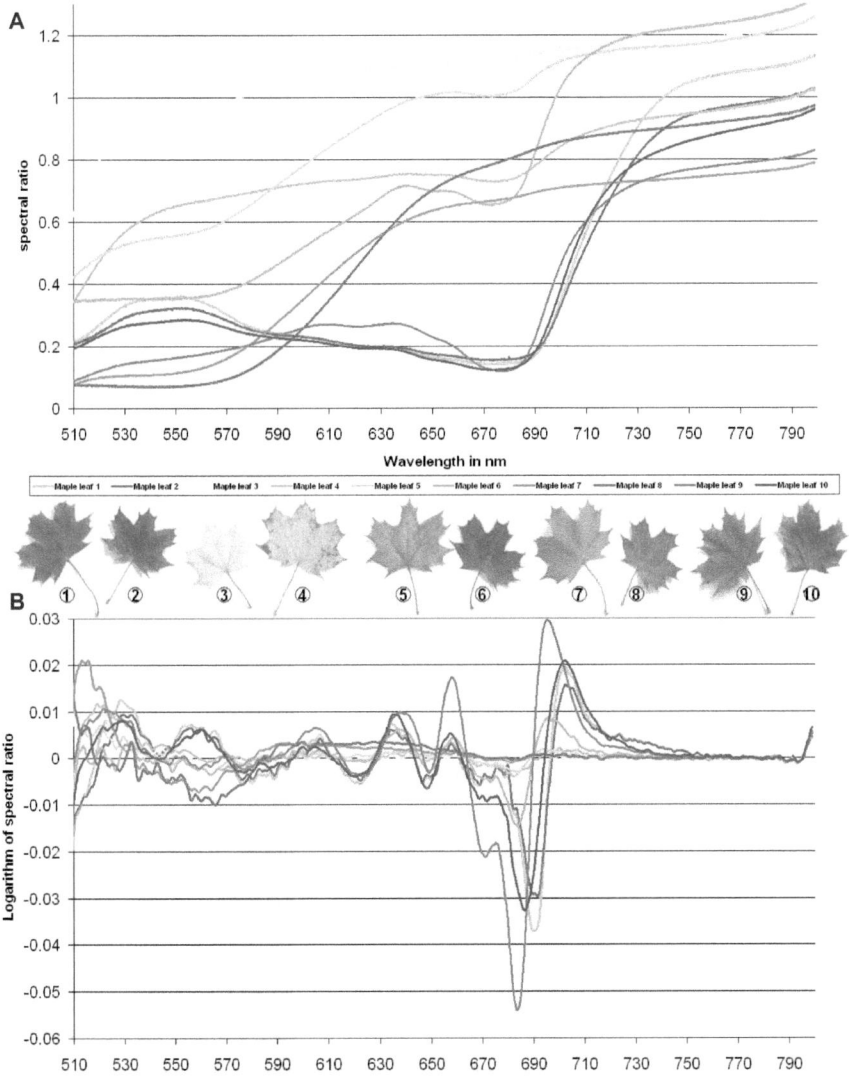

Figure 8.5
Spectral reflectance of maple leaves in autumn. A) shows the original, B) the highpass-filtered signal.
The leaves 1 to 8 are ordered roughly according to colour progression over time. First the absorption decreases in the red and green wavelength-ranges, increasing the reflectance, producing bright yellow leaves. Yellow is the colour derived when adding strong red and green light. After that, pigments are built up absorbing the blue and green wavelengths, producing orange and red colours.
Leaf 9 appears to display several stages in parallel. Leaf 10 was taken from a branch almost completely broken off from the tree, drying out.

8.6 Lichen

Lichen are also interesting to measure. They exist in all vegetation zones, sometimes growing on top of the leaves of trees, e.g. in the tropic rainforest, as well ws on branches, tree trunks and even stones. In polar regions and in the mountains beyond the tree-line they may be the single most important plant.

The signal of lichen is usually not easy to register from satellites because in most vegetation-zones it is usually blocked by the stronger signal of broadleaf or conifer vegetation. When they are growing on tree branches and trunks, leaves shields their signal. But the signal might become visible seasonally in areas where leaves are shed.

Lichen are a symbiotic life form of mosses and algae. They possess the rare capability of being able to survive total dry-out. As soon as moisture is available again (e.g. morning dew, rain etc.) they revive. To see the different signal lichen produce under dry and moist conditions, we first measured lichen attached to a branch after several weeks of drying out. Those signals are shown as yellow and light-grey in Figure 8.6. The upper Figure shows the general reflectance signal and the lower the high-frequency structures of the signal. After the dry measurements the whole branch was exposed to direct contact with water and afterwards stored at 100% air moisture for a full day. Then measurements were repeated. Those signals are depicted in orange and green. The first apparent difference is that the reflectance of the moist lichen is in general massively reduced showing the increase in absorption inside the reviving lichen. In Figure 8.7.2 the increased absorption, especially in the chlorophyll, is very apparent for both types of lichen. There is not much change in absorption in the wavelength ranges of 550 to 620 nm but there is a very strong strengthening of reflectance in the 520 to 540 nm range. This might be explained by the now apparent colour change which in the grey lichen is tending towards the green and in the yellow lichen is tending towards orange. These measurements were done to show the capabilities of DOAS-measurements for the monitoring of lichen.

For satellite-applications, the lichen of the Tundra are especially interesting since they dominate the vegetation in those areas. Unfortunately I was not able to access and measure typical Tundra lichen, therefore a reference-spectrum for DOAS was not produced.

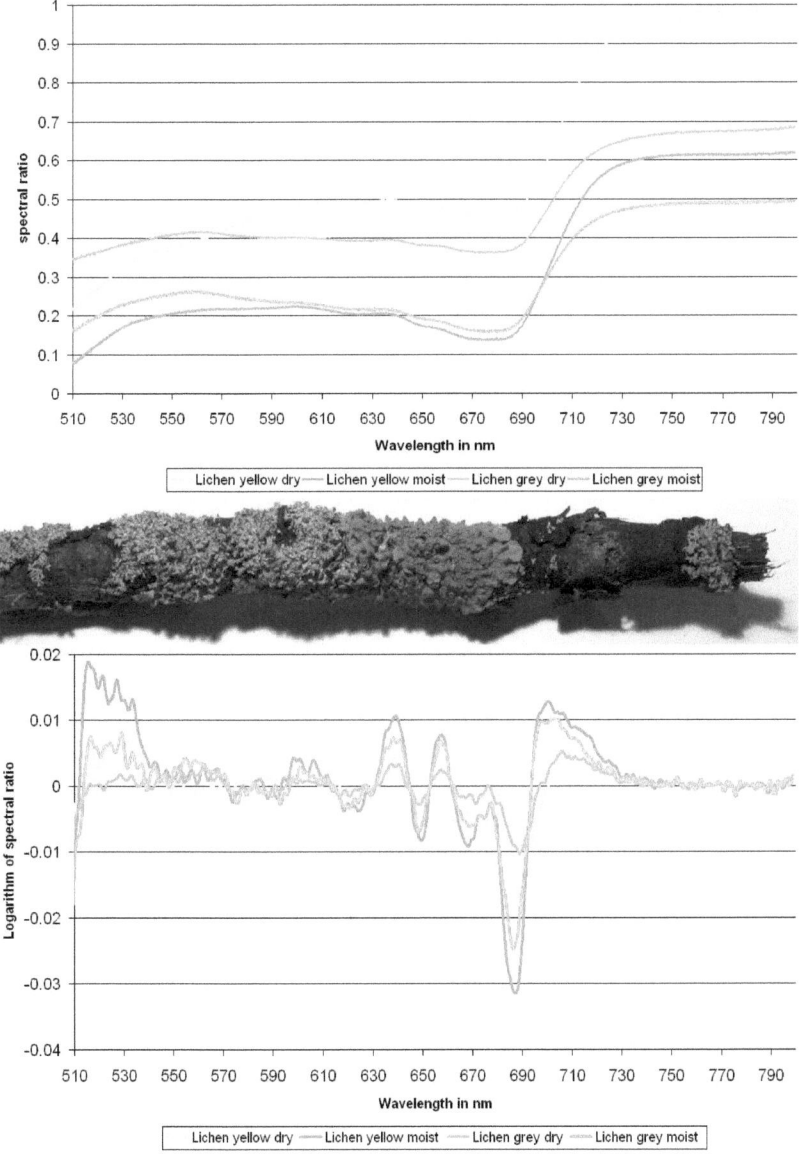

Figure 8.6
Measuring two different species of lichen in dry and moist condition. Spectral absorption intensifies especially from about 600 nm onwards after moisturizing.

8.7 Influence of incidence angles

In this Chapter features changing with the incidence angle of illumination and of the measurement instrument are studied. Figure 8.7.1 shows the definition of the incidence angle. 0°-incidence angle is nadir, completely in the vertical. The difference from the vertical is then called the incidence angle. We considered changes of the incidence angle separately for light source and for the instrument. Only one of the angles was changed, whereas the other remained in the nadir position.

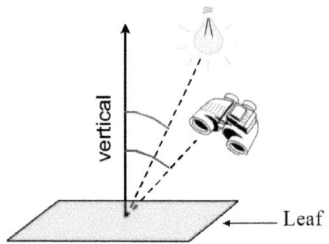

Figure 8.7.1
The incidence angle is defined as the deviation from the vertical (i.e. nadir). In this case the plane of reference is the leaf.

These investigations were done over several leaves. The response, especially in the near-infrared, was always strong. The structure of the leaf-build up definitely has an influence on the reflectance capabilities. This is similar to the structural colours of e.g. insects (see e.g. Brydegaard et al., 2009 or Parker et al., 2005). Structural colours are produced by regular three-dimensional structures, e.g. on the wings of insects. However, unlike to the case of insects, where these structures interfere with visible light, the effects of leaves may be more pronounced in the near-infrared.

The first example shows the effects over a Maple leaf. In separate Figures changes in illumination angle and in viewing angle (i.e. view of the measurement instrument) are shown. In both cases changes are minor for wavelengths up to 650 nm. Towards higher wavelengths the spectral structures become variable. The alterations are different for changes of illumination and of viewing angle. This applies also to measurements over other leaves. These changes from 650 nm on came as a surprise. This is still in the visible spectrum. Literature suggested these kinds of influences arise only from about 750 nm in the near-infrared (see Chapter 2.2.2 or e.g. Barrett & Curtis, 1992).

Another example for the changes of reflectance is a Plum leaf, although the changes are not the same as over a Maple leaf, which suggests that the different structures of Maple and Plum leaf do have an influence. Both measurements show that e.g. the absorption band of chlorophyll is more influenced by changing angles of the viewing instrument. Changing illumination angles show almost no difference at all at the chlorophyll a absorption band. In both cases, changes of the viewing angle have stronger effects also on the lower wavelength ranges between 650 and 680 nm, where both show more intensive shifts of absorption maxima than occur with changes of illumination angle.

In all cases the changes in the near-infrared are most pronounced. This shows clearly the effect of angle changes because in the measurements before we saw that the area of about 750 nm and more is relatively low in structures. This changes decidedly when angles are altered.

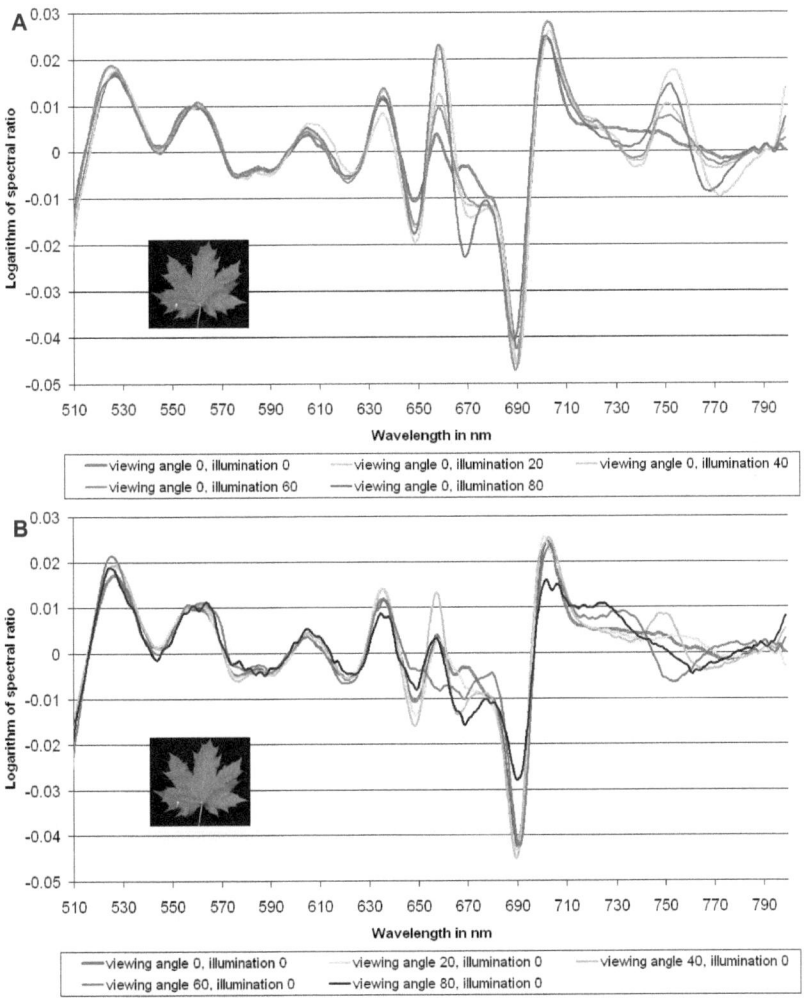

Figure 8.7.2
Effect of changing illumination and viewing angle on high-frequency reflection structures of a maple leaf.
A) shows the effect of changing illumination angle. B) shows the effect of changing viewing angle.

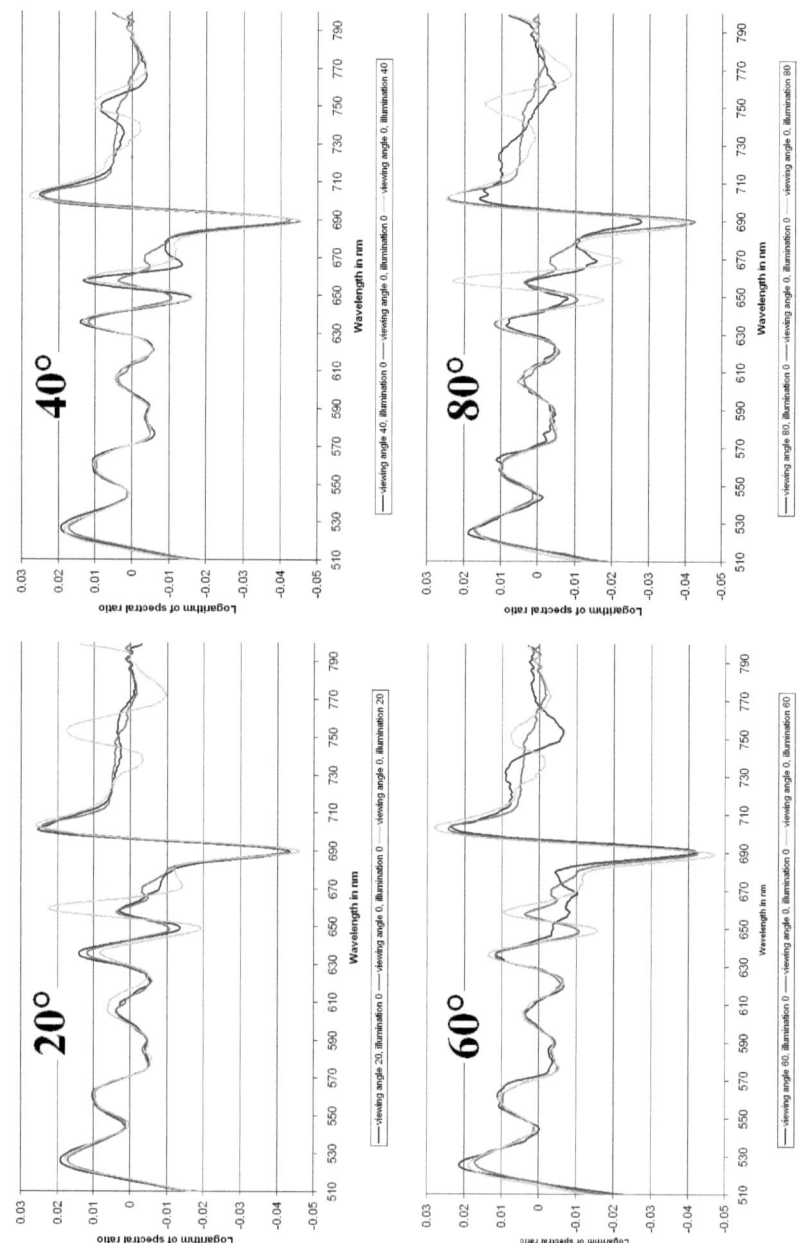

Figure 8.7.3
Effect of changing illumination and viewing angle on reflection of maple leaf. Seperated into comparison of same angle against all nadir measurement. For all graphs: red line all nadir, blue: altered viewing angle, green: altered illumination angle. Changes are mostly at similar wavelengths, but quite different in quality and amplitude.

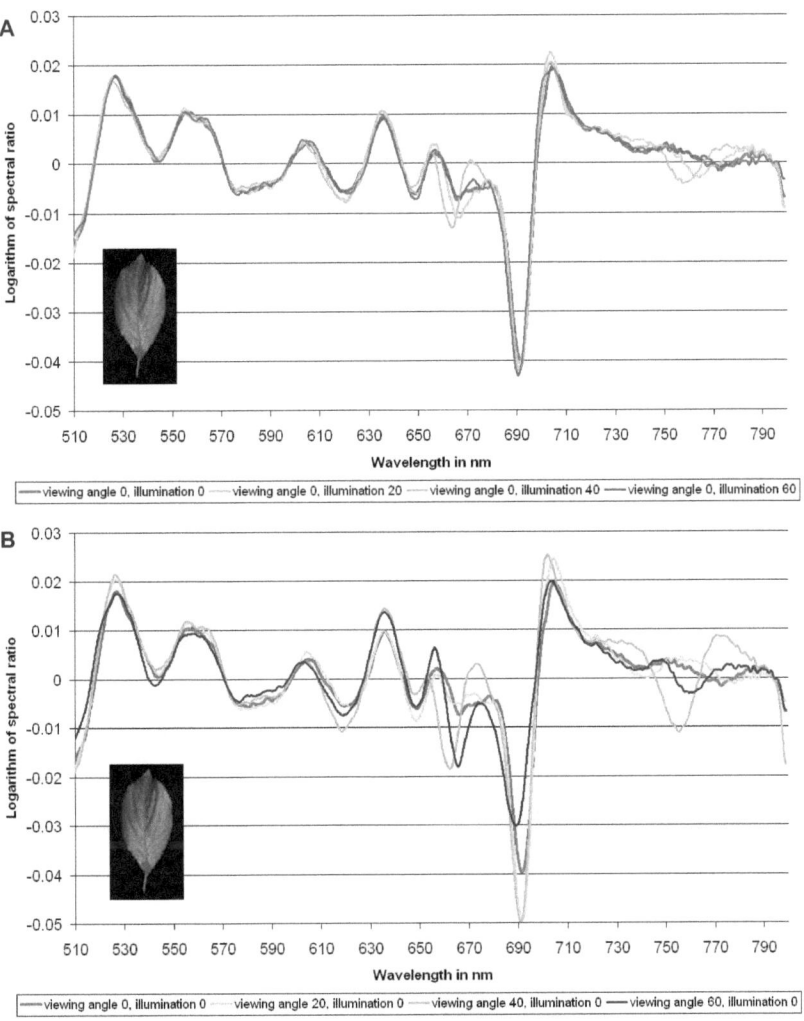

Figure 8.7.4
Effect of changing illumination and viewing angle on reflectance of a plum leaf. A) shows the effect of changing illumination angle. B) shows the effect of changing viewing angle.

These measurements of changes were also done for several other broadleaf leaves (see Appendix B) and over Maize, Maize being the only grass in this investigation. However, the same general changes apply for Maize as for the broadleaf trees. Changes start decidedly around 650 nm upwards and are extremely strong in the near-infrared, so there is no general difference between broadleaf leaves and grass leaves. This is a very important finding for satellite measurements. Trees usually direct their leaves to optimum availability for light. They are mostly vertical to the incoming light. But even the broadleaf trees do not always have all their leaves vertical to the sun, but show all kinds of angles towards the sun. Several of these features of shifting angles of illumination and also towards the satellite instrument should be visible at the same time. Even more important are these findings for the grasses, because grasses often shoot their leaves upwards, but when they grow bigger they start folding over. Then a single leave has progressive viewing angles from vertical to the sun to almost horizontal to the sun. All the differences in illumination and viewing angle are expected to be observed in parallel. This makes satellite based DOAS-retrievals of vegetation quite complicated.

A positive finding for satellite retrievals is that the changing angle of the light source has less influence than the changing angle of the instrument. This is fortunate, since the local solar zenith angle is strongly dependent on daytime, season and latitude, meaning it is constantly changing during every orbit and between orbits. The viewing-angle perpendicular to the track is quite small for SCIAMACHY, so that this effect should remain minor. These effects were therefore ignored during first satellite retrievals, although this topic merits further research.

To minimise influences of angular dependence, the Mini-MAX-DOAS was always at nadir, while the artificial light-source was always as close to nadir as possible.

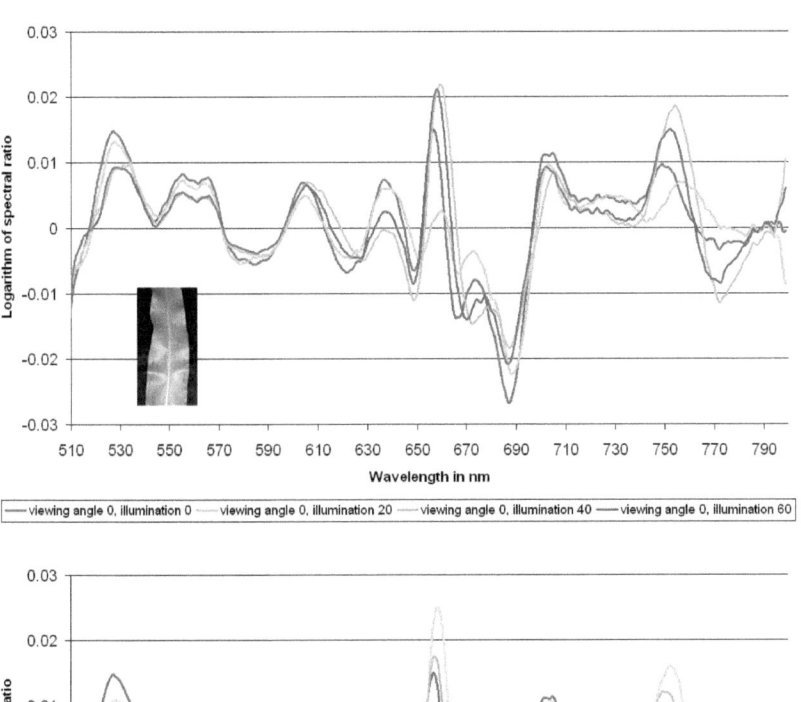

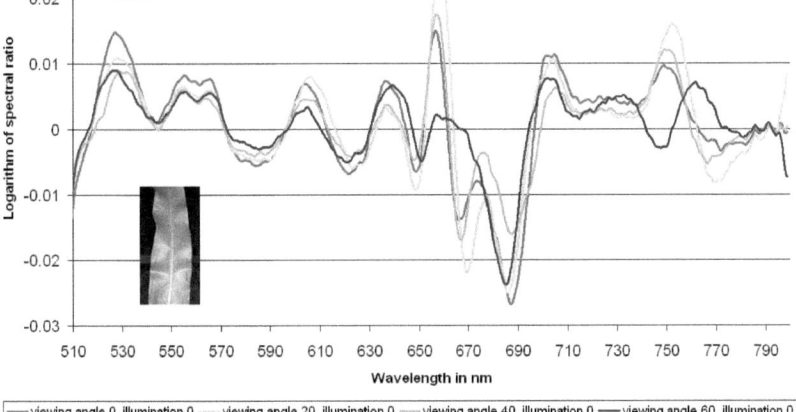

Figure 8.7.5
Effect of changing illumination and viewing angle on a maize leaf. The red line shows the reflection for both angles nadir. Changing illumination is shown in green colours, changing viewing angle in blue. Due to the growing characteristics of sweet-grass leaves, all angles may apply to every single leaf.

8.8 Summary

The results presented in this Chapter prove the capabilities of a Mini-MAX-DOAS instrument for research on different characteristics of vegetation. This is a perspective for future use in research.

Important results for the measurement of reference spectra are that:
- the instrument should be at nadir and the light-source as close as possible to nadir to avoid effects as described in Chapter 8.7.
- when measuring coniferous plants, the needles should remain attached to the branch.

Interesting results for considerations when applying the new reference-data to satellites are that:
- the viewing angle of the instrument has considerable influence on vegetation-reflectance. Therefore the used satellites should view at almost nadir and the opening angle should not be too large. To define the acceptable limit, further research is required.
- changing angles of incident sunlight have an effect on vegetation-reflectance. A cut-off at high SZA should be considered. To define this borderline case, further research is necessary.
- the strength of grass-signals changes considerably over the vegetation period. Therefore the measured signal should be combined with season information before interpreting amount of biomass.
- deciduous trees change their absorption signal significantly at the onset of autumn. To monitor this effect, different reference spectra are needed.
- upper and lower side of leaves have different reflection-characteristics. Although the upper side is mostly turned towards the satellite, wind may alter that. How strongly this will alter the signal received at the satellite should be further investigated.

9 Optimizing DOAS-fit for vegetation spectra

9.1 Choice of parameters for the spectral analysis

For the analysis of the GOME spectra (see Chapter 7), a wavelength range between 615 and 681 nm was used. This choice was motivated by the wavelength range of the respective spectral channel (at the lower wavelength edge) and the strong absorption lines of H_2O and O_2 (at the upper wavelength edge).

The first analysis using the new vegetation reference spectra, the original wavelength range was chosen. An example of the spectral fitting process for a measurement over rain forest is shown in Figure 9.1.1.

While the atmospheric absorptions of H_2O and O_2 can be analysed with high accuracy, the vegetation signal is rather low, especially compared to the magnitude of the fit residual. It is obvious that systematic structures of the spectral residual are caused by an imperfect fit of the H_2O and O_2 absorption spectra. This indicates that the available absorption spectra found in the literature (e.g. from the HITRAN data base) are not sufficiently representative for the atmospheric absorptions.

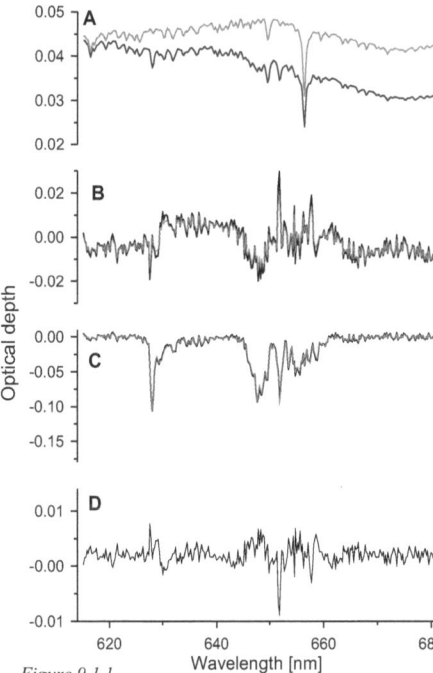

In order to make use of a broader range of the vegetation spectra, in the next step the wavelength range used for the analysis was extended. While the lower wavelength edge is fixed at about 615nm due to the instrumental properties, the upper wavelength edge was extended to 712nm leading to a wavelength interval which is about 30nm larger than the original fitting range. The new wavelength range in particular includes the strong spectral features of the vegetation reference spectra around 690nm. The upper limit of the wavelength range was not further extended because of the strong water vapour absorption above 712nm. Unfortunately, also parts of the wavelength range had to be excluded (e.g. around 685 – 702nm) because of the strong O_2 and H_2O absorptions at these wavelengths. An additional small interval was excluded around 651nm. An example of the spectral fitting process for the same measurement as in Figure 9.1.1 is shown in Figure 9.1.2.

Figure 9.1.1
Fit result for a measurement over Rain forest. A: solar spectrum (blue) and atmospheric measurement (red); B: sum of all vegetation spectra; C: sum of the H_2O and O_2 reference spectra; D: Residual. As vegetation reference spectra Betulaceae, Larix, and Pinus were included.

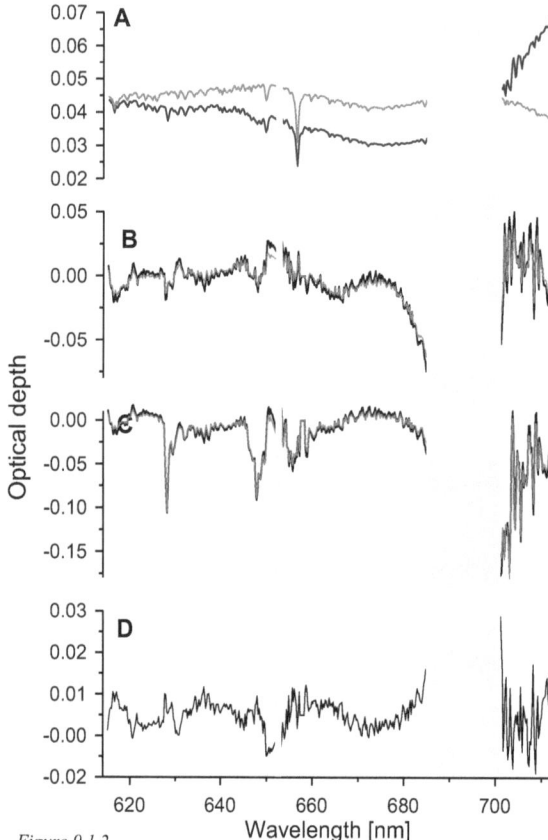

On the one hand, more spectral information of the vegetation spectra could now be considered in the spectral analysis. However, in the end the fit quality is not satisfactory, because with the larger wavelength range, also the fit residual strongly increased. Still the vegetation signal can not be clearly seen in the fit results. Also the largest part of the strong vegetation signature around 690nm had to be excluded from the fit.

To improve the analysis, it was decided that besides the H_2O and O_2 absorption spectra from laboratory measurements (from the HITRAN data base), also reference spectra obtained from the satellite measurements themselves should be included in the spectral analysis. Such spectra should largely improve the fit quality because a) exactly the same spectral properties as for the other measured spectra are used, and b) the O_2 and H_2O absorption are observed directly in the atmosphere.

Figure 9.1.2
Fit result for a measurement over Rain forest. A: solar spectrum (blue) and atmospheric measurement (red); B: sum of all vegetation spectra; C: sum of the H_2O and O_2 reference spectra; D: Residual. As vegetation reference spectra Betulaceae, Larix, and Pinus were included. The green rectangles indicate the gaps in the fitting window.

The determination of well suited reference spectra from the satellite observations is not a trivial task. The following conditions have to be considered:
a) measured spectra have to taken over areas without vegetation at the ground
b) always two measured spectra have to be divided by each other to eliminate the strong Fraunhofer lines. The absorptions of H_2O and O_2 in both spectra should differ, while other spectral properties should be similar.
c) measurements at different atmospheric absorptions (both for O_2 and H_2O) have to be chosen to consider effects of spectral saturation.

Based on these requirements observations at different latitudes and for different cloud altitudes were selected. All determined absorption spectra simultaneously contain O_2 and H_2O absorption lines, but with different strengths. In total three different atmospheric reference spectra for O_2 and H_2O were included in the spectral fit (in addition to the reference spectra measured in the laboratory). An overview on all O_2 and H_2O reference spectra finally used in the spectral analysis is shown in Figure 9.1.3.

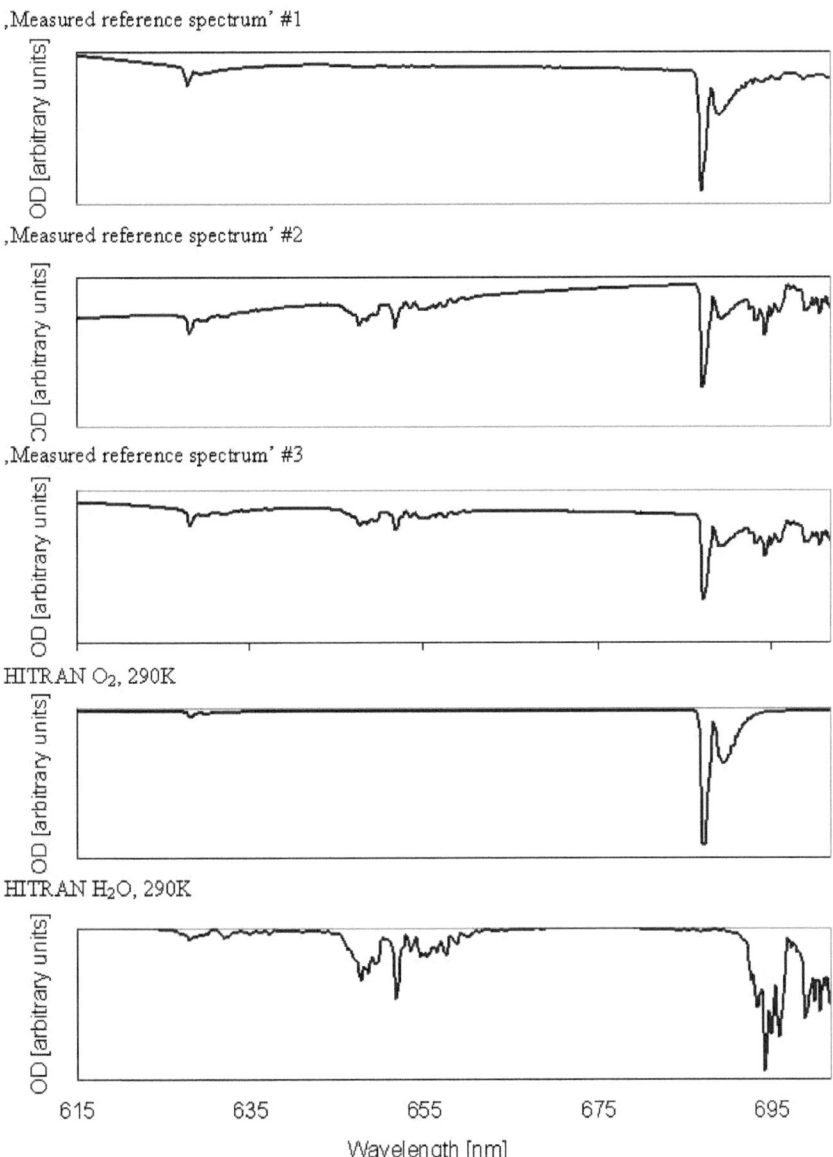

Figure 9.1.3
Overview on the different reference spectra for O_2 and H_2O used in the spectral analysis. The upper three reference spectra were obtained from the satellite measurements themselves (see also text); the two lower spectra were measured in the laboratory.

An example of the spectral fitting process for the same measurement as in Figure 9.1.1 is shown in Figure 9.1.4. Compared to the previous examples, the residual structures are smaller indicating that the new reference spectra can well describe the atmospheric absorptions in the measured spectrum. In addition, also the remaining gaps are smaller than in Figure 9.1.3, and most of the strong vegetation reflectance feature at 690nm is well covered by the spectral analysis. Compared to the previous examples, the vegetation reflectance signal is now well visible.

These new settings were used in the following to retrieve the vegetation reflectance signals from the spectra measured by the SCIAMACHY instrument.

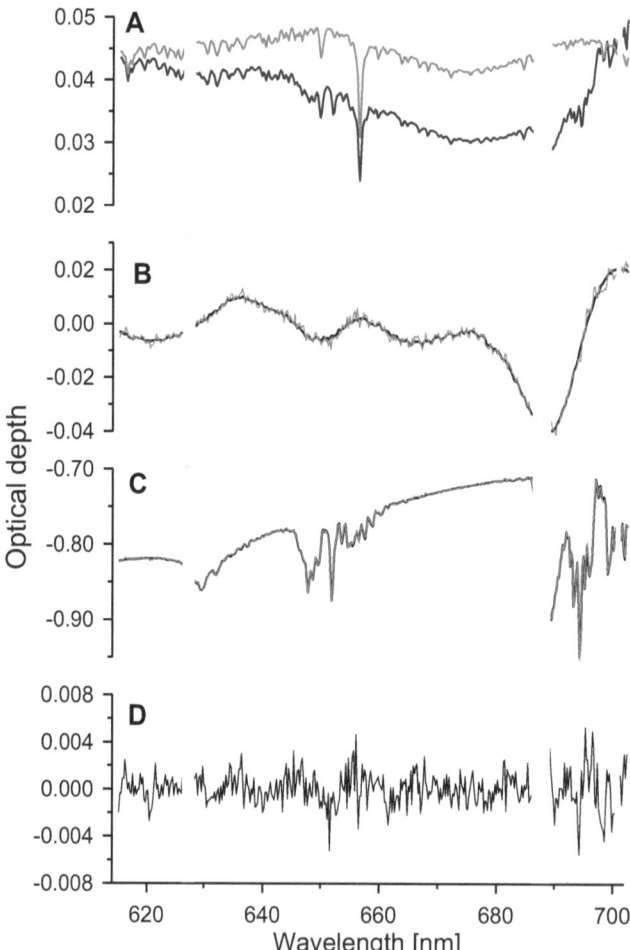

Figure 9.1.4
Fit result for a measurement over Rain forest. A: solar spectrum (blue) and atmospheric measurement (red); B: sum of all vegetation spectra; C: sum of the H_2O and O_2 reference spectra; D: Residual. As vegetation reference spectra Betulaceae, Larix, and Pinus were included. The green rectangles indicate the gaps in the fitting window.

9.2 Reference areas

To test the new the DOAS-fit, was tested over areas with known vegetation. Within these areas vegetation has to be rather uniform over large areas according to the spatial satellite resolution. To identify these areas, "Alexander Weltatlas" (Schulze 1989) and the ESA "Globcover 300m" map (see Chapter 4) were used.

Only cloud-free measurements were chosen, as determined based on the HICRU algorithm, which is described in Chapter 4. The selected areas were chosen based on the Globcover project.

All reference areas were chosen to with a minimum extension of 1° north to south and 2° west to east, allowing the area to be large enough to contain several satellite-pixels over the same class of vegetation.

Nine reference areas were defined to test the new reference spectra over different vegetation classes under cloud-free conditions. Dates, locations and common plant-societies are shown in Table 9.1.

Table 9.1: List of reference areas, giving dates of cloud-free conditions, vegetation class and geographic degrees of boundaries.

Date	Vegetation	Region	Latitude	Longitude	Common plant-societies
01.06.05	Rain forest	SE Manaus	-4.5° to -6°	-59° to -57°	Intensively mixed broadleaf.
15.06.05	Grass-land	Botswana	-21° to -24°	-20° to 25°	Dominating sweet-grasses. Towards NW mixing in of deciduous signal.
20.06.05	Taiga	S Yakutsk	61° to 59°	126° to 130°	Conifers dominated by Larix, but mixing in of Pinaceae and Betulaceae. Grass-signatures expected in spaces between trees.
25.06.05	Tundra	N Workuta	69.5° to 68°	61° to 67°	Lichen, sedge (Carex), dwarf-trees and smaller forms of Pinaceae, Betulaceae and Rosaceae
29.06.05	Rain forest	W Manaus	-2° to -3°	-64° to -62°	Intensively mixed broadleaf.
05.07.05	Taiga	S Yakutsk	61° to 59°	126° to 130°	Conifers dominated by Larix, but mixing in of Pinaceae and Betulaceae. Grass-signatures expected in spaces between trees.
09.07.05	Coniferous forest	SW Hudson Bay	55° to 51°	-99° to -90°	Conifers dominated by Spruce (Pinaceae), but mixing in of other Pinaceae, Larix, Betulaceae, Aceraceae and Rosaceae expected. Grass-signatures likely.
09.07.05	Rain forest	SE Manaus	-4.5° to -6°	-59° to -57°	Intensively mixed broadleaf.
14.07.05	Tundra	N Workuta	69.5° to 68°	61° to 67°	Lichen, sedge (Carex), dwarf-trees and smaller forms of Pinaceae, Betulaceae and

Date	Class	Location	Latitude	Longitude	Description
					Rosaceae
29.07.05	Agriculture = grass-land	SE Chicago	42° to 39°	-87° to -82°	Dominating sweet-grasses. Mostly annuals, north-western part also
06.08.05	Rain forest	W Manaus	-2° to -3°	-64° to -62	Intensively mixed broadleaf.
24.08.05	Grass-land	Botswana	-21° to -24°	20° to 25°	Dominating sweet-grasses. Towards NW mixing in of deciduous signal.
31.08.05	Tundra	N Workuta	69.5° to 68°	61° 67°	Lichen, sedge (Carex), dwarf-trees and smaller forms of Pinaceae, Betulaceae and Rosaceae
06.09.05	Agriculture = grass-land	SE Chicago	42° to 39°	-87° to -82°	Dominating sweet-grasses. Mostly annuals, north-western part also perennials. Some deciduous.
06.09.05	Deciduous forest	SW Pittsburgh	40° to 38.5°	-82° to .80°	Well mixed deciduous families, Grasslands (possibly sweet-grasses) in the valleys
07.09.05	Deciduous forest	SW Pittsburgh	40° to 38.5°	-82° to .80°	Well mixed deciduous families, Grasslands (possibly sweet-grasses) in the valleys
01.10.05	Agriculture = grass-land	SE Chicago	42° to 39°	-87° to -82°	Dominating sweet-grasses. Mostly annuals, north-western part also perennials. Some deciduous.
17.02.06	Agriculture = grass-land	W SantaFe	-31° to 33°	-64° to -61°	Dominating sweet-grasses. Western part mostly wheat, eastern part mostly maize.
05.03.06	Agriculture = grass-land	W SantaFe	-31° to 33°	-64° to -61°	Dominating sweet-grasses. Western part mostly wheat, eastern part mostly maize.
31.03.06	Grass-land	Botswana	-21° to -24°	20° to 25°	Dominating sweet-grasses. Towards NW mixing in of deciduous signal.
13.04.06	Grass-land	Botswana	-21° to -24°	20° to 25°	Dominating sweet-grasses. Towards NW mixing in of deciduous signal.

The Globcover-classes within a reference area were determined according to the Globcover project. The respective classes are described in Tables 9.2 and 9.3.

Table 9.2: Vegetation-classes of the Globcover 300m map. See Globcover map Figures 4.3.1 and 4.3.2.

1	Cultivated and Managed areas / Rainfed cropland
2	Post-flooding or irrigated croplands
3	Mosaic cropland (50-70%) / vegetation (grassland/shrubland/forest) (20-50%)
4	Mosaic vegetation (grassland/shrubland/forest) (50-70%) / cropland (20-50%)
5	Closed to open (>15%) broadleaved evergreen and/or semi-deciduous forest (>5m)
6	Closed (>40%) broadleaved deciduous forest (>5m)
7	Open (15-40%) broadleaved deciduous forest/woodland (>5%)
8	Closed (>40%) needle-leaved evergreen forest (>5m)
9	Closed (>40%) needle-leaved deciduous forest (>5m)
10	Open (15-40%) needle-leaved deciduous or evergreen forest (>5%)
11	Closed to open (>15%) mixed broadleaved and needle-leaved forest
12	Mosaic forest or shrubland (50-70%) and grassland (20-50%)
13	Mosaic grassland (50-70%) and forest or shrubland (20-50%)
14	Closed to open (>15%) shrubland (<5m)
15	Closed to open (>15%) grassland
16	Sparse (<15%) vegetation
17	Closed (>40%) broadleaved forest regularly flooded, fresh water
18	Closed (>40%) broadleaved semi-deciduous and/or evergreen forest regularly flooded, saline water
19	Closed to open (>15%) grassland or shrubland or woody vegetation on regularly flooded or waterlogged soil, fresh, brakish or saline water
20	Artificial surfaces and associated areas (Urban areas >50%)
21	Bare areas
22	Water bodies
23	Permanent Snow and Ice
24	No data

Table 9.3: Relation of vegetation classes in this thesis to Globcover classes.

Vegetation	Globcover 300m
Rainforest	Closed to open (>15%) broadleaf evergreen and / or semi-deciduous forest (>5m)
Deciduous forest	Closed (>40%) broadleaf deciduous forest (>5m)
Conifer forest / Taiga	Closed (>40%) needle-leaved evergreen forest (>5m)
Tundra	Sparse (<15%) vegetation
Grassland: agriculture	Cultivated and Managed areas / Rainfed cropland
Grassland: unmanaged / arid	Closed to open (>15%) grassland

9.2.1 Selected reference areas

The specified reference areas are introduced here, showing their location and land-use in maps from "Alexander Weltatlas" (Schulze 1989), extracts from the Globcover map and the HICRU-results for the cloud-free days.

9.2.1.1 Rainforest

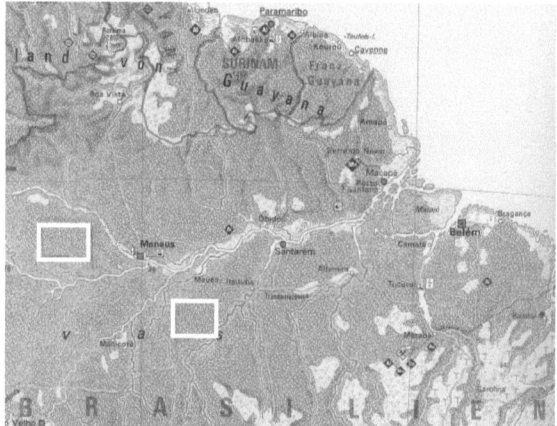

Figure 9.2.1
Land-use map for the reference areas rainforest. Excerpt from Alexander Weltatlas (Schulze, 1989)

Two areas are within the Amazon rainforest, one just west of Manaus, the other south-east of Manaus:

The reference area WManaus is between the Rivers Rio Negro and Amazon. The rainforest in this area is still quite pristine. The Globcover map shows a very even signal for broadleaf evergreen forest.

The reference area SEManaus is between the Rivers Madeira and Tapajos, two southern tributaries to the Amazon river. This area also shows a very even signal for broadleaf evergreen forest in the Globcover map. It is also close to invading logging activities and might display signal-changes as a result to that effect in the near future.

For both areas the response for broadleaf families should be strong, but also southern pine might contribute to the signal. Grasses should not respond.

Figure 9.2.2
Vegetation map for reference area W Manaus. Excerpt from Globcover map.

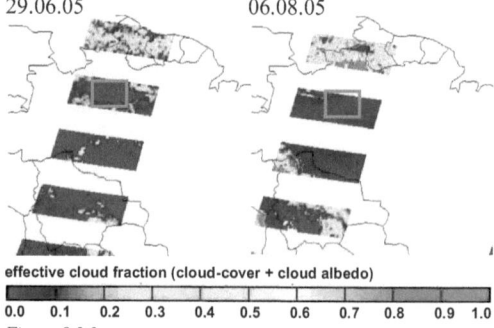

Figure 9.2.3
Cloud-cover analysed with HICRU (Grzegorski, 2009) for days without clouds over reference area

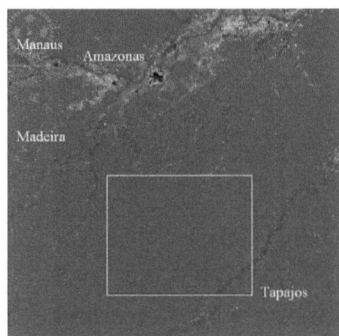
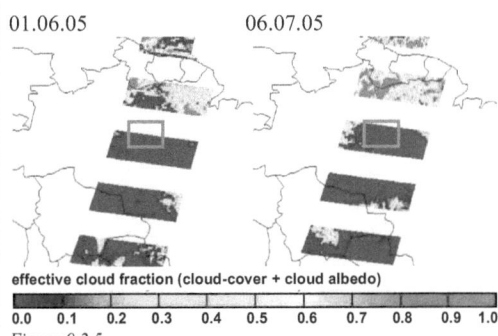

Figure 9.2.4
Vegetation map for reference area SE Manaus. Excerpt from Globcover map.

Figure 9.2.5
Cloud-cover analysed with HICRU (Grzegorski, 2009) for days without clouds over reference area.

9.2.1.2 Deciduous forest

The reference area SPittsburgh is situated just west of the Appalachian Mountains, in the plains of the River Ohio.

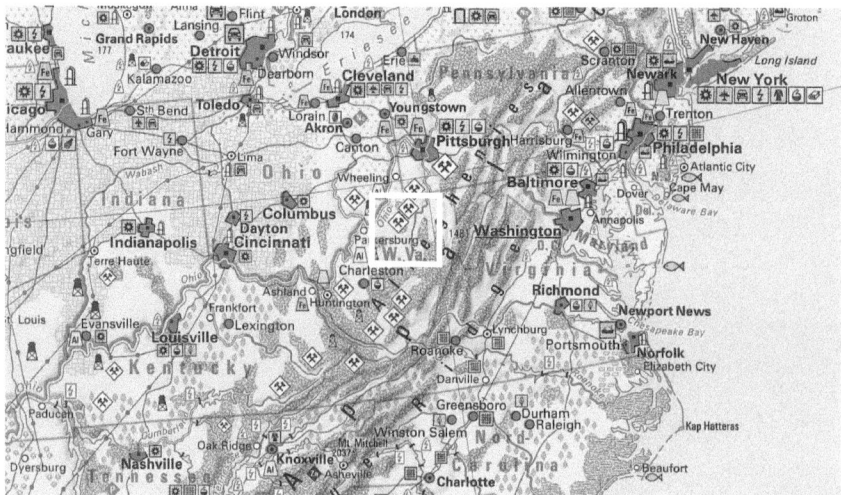

Figure 9.2.6
Land-use map for the reference area deciduous forest. Excerpt from Alexander Weltatlas (Schulze, 1989)

According to Globcover the dominating vegetation is broadleaf deciduous forest of more then 40% cover. This means that deciduous classes should respond strongly, but grasses should also respond, especially sweet-grasses owing to intermittent pasture and agriculture.

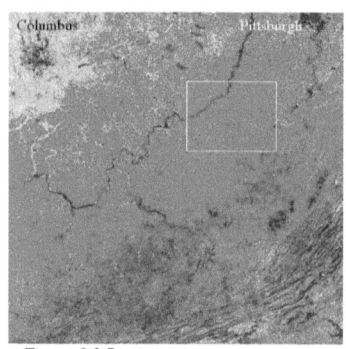

Figure 9.2.7
Vegetation map for reference area SW Pittsburgh.. Excerpt from Globcover map (©ESA)

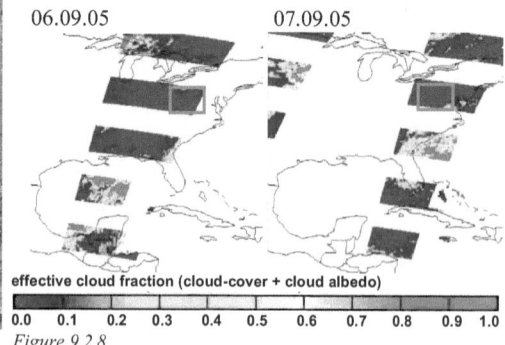

Figure 9.2.8
Cloud-cover analysed with HICRU (Grzegorski, 2009) for days without clouds over reference area. Darkest blue virtually cloud-free, red full cloud-cover.

9.2.1.3 Coniferous forest / Taiga

Large areas of uninterrupted coniferous forests are mostly restricted to the northern latitudes of America and Eurasia. The two areas are on two different continents. One is in Canada, south-west of the Hudson Bay. The other is in Siberia, south-east of Yakutsk.

Figure 9.2.9
Land-use map for the reference area coniferous forest / Taiga. Excerpt from Alexander Weltatlas (Schulze, 1989)

The reference area SWHudsonBay shows dominating needle-leaved evergreen forest of more than 40% cover intermixed with mosaic shrub- and grassland. The presence of many waterbodies in the area suggests that grass in this case should be dominated by sedge (Carex). The shrubs will add some broadleaf contribution. Most of the vegetation signal should be contributed by conifers, in this area especially from spruce.

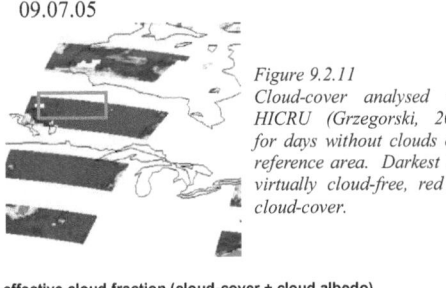

Figure 9.2.11
Cloud-cover analysed with HICRU (Grzegorski, 2009) for days without clouds over reference area. Darkest blue virtually cloud-free, red full cloud-cover.

Figure 9.2.10
Vegetation map for reference area SW Hudson Bay. Excerpt from Globcover map. (©ESA)

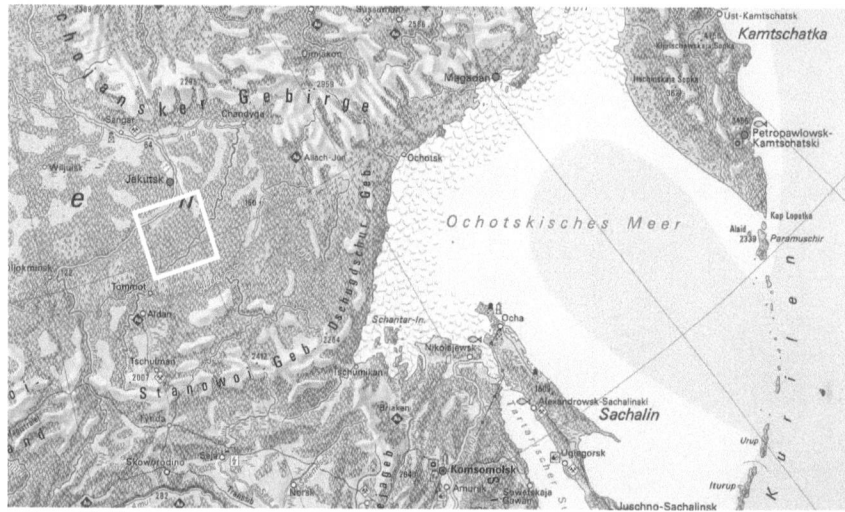

Figure 9.2.12
Land-use map for the reference area coniferous forest / Taiga. Excerpt from Alexander Weltatlas (Schulze, 1989)

The coniferous forest in the reference area SYakutsk shows a much more uniform classification of needle leaved deciduous or evergreen forest of more then 40% cover between the Rivers Alden and Lena. Coniferous signal should be rather strong here, especially Larix.

Northern coniferous forests rarely have closed crown-cover as might happen in more moderate climates, so there will always be a mixed signal from vegetation growing beneath the conifers. Sedge (Carex) is to be expected as well as some broadleaf (especially Betulaceae and Rosaceae).

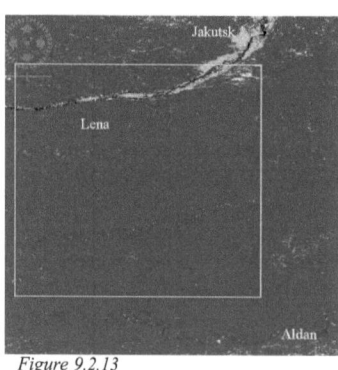

Figure 9.2.13
Vegetation map for reference area S Yakutsk. Excerpt from Globcover map (©ESA)

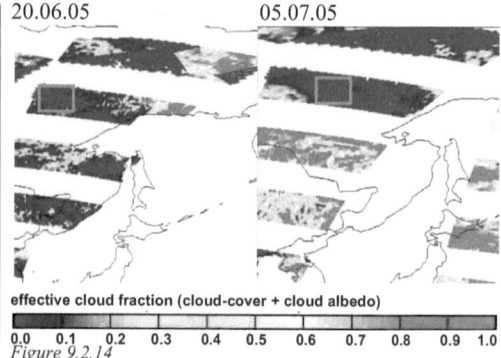

Figure 9.2.14
Cloud-cover analysed with HICRU (Grzegorski, 2009) for days without clouds over reference area. Darkest blue virtually cloud-free, red full cloud-cover.

9.2.1.4 Tundra

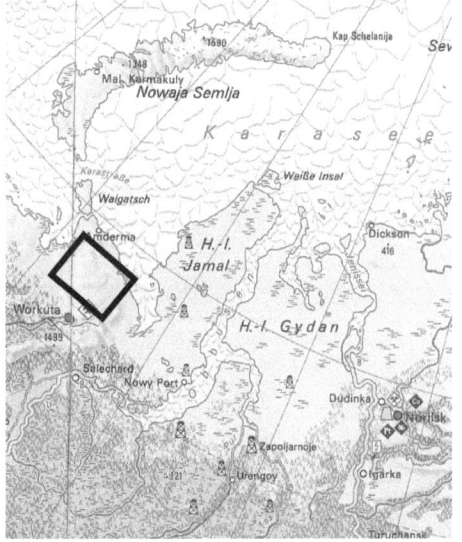

The reference area for Tundra is in Siberia, between Workuta and the Kara Sea.

The Tundra NWorkuta shows in the Globcover map as mostly less than 15% vegetation intermixed with small areas of mosaic shrub- and grassland. Vegetation here is dominated by lichen. Intermittent sedge (Carex) and dwarf trees (especially Betulaceae and Rosaceae) might grow alongside these during the arctic summer.

Figure 9.2.15
Land-use map for the reference area Tundra. Excerpt from Alexander Weltatlas (Schulze, 1989)

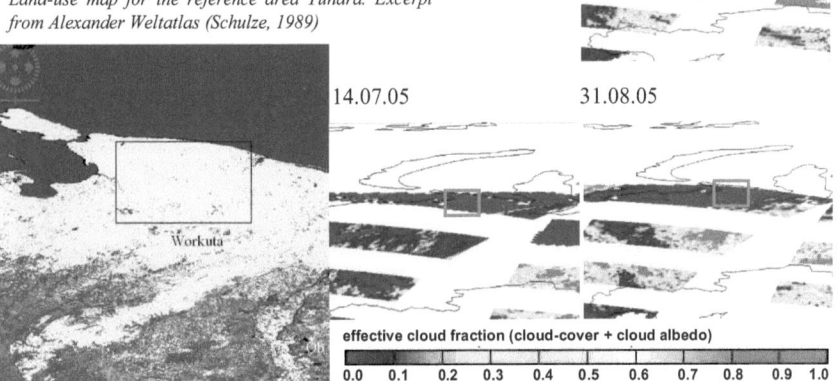

Figure 9.2.16
Vegetation map for reference area N Workuta. Excerpt from Globcover map (©ESA)

Figure 9.2.17
Cloud-cover analysed with HICRU (Grzegorski, 2009) for days without clouds over reference area. Darkest blue virtually cloud-free, red full cloud-cover.

111

9.2.1.5 Grassland: agriculture

For agricultural land we chose the large areas in Argentina between Cordoba and Rosario.

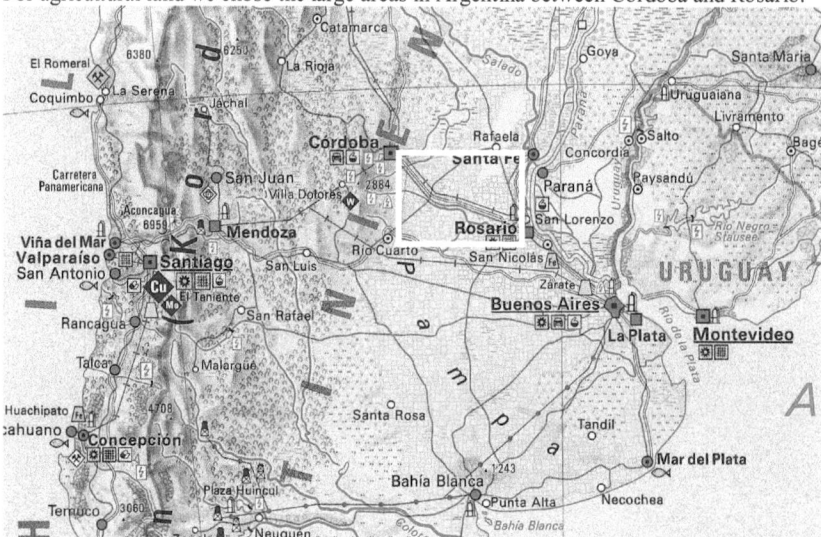

Figure 9.2.18
Land-use map for the reference area grassland / agriculture. Excerpt from Alexander Weltatlas (Schulze, 1989)

The reference area WSantaFe spans much of the agricultural areas in Argentina. According to the Globcover map this area is mostly covered with rainfed cropland mixed in occasionally with some shrub-land.

The land-use map from Schulze (1989) suggests dominating wheat in the west and dominating maize in the east of the area. Both are sweet-grasses.

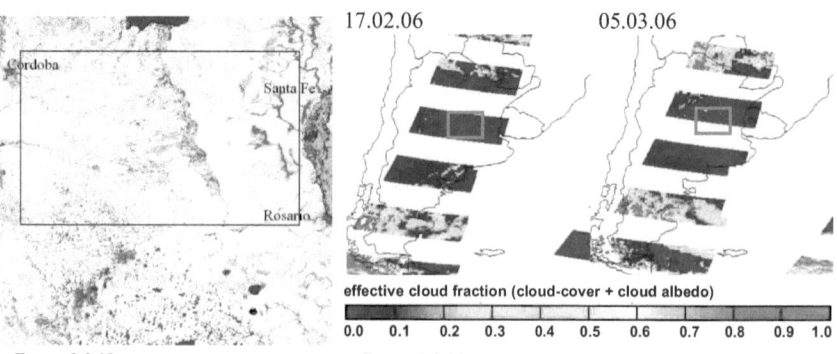

Figure 9.2.19
Vegetation map for reference area SE Cordoba. Excerpt from Globcover map (©ESA)

Figure 9.2.20
Cloud-cover analysed with HICRU (Grzegorski, 2009) for days without clouds over reference area. Darkest blue virtually cloud-free, red full cloud-cover.

112

Another reference area for agriculture is SEChicago.

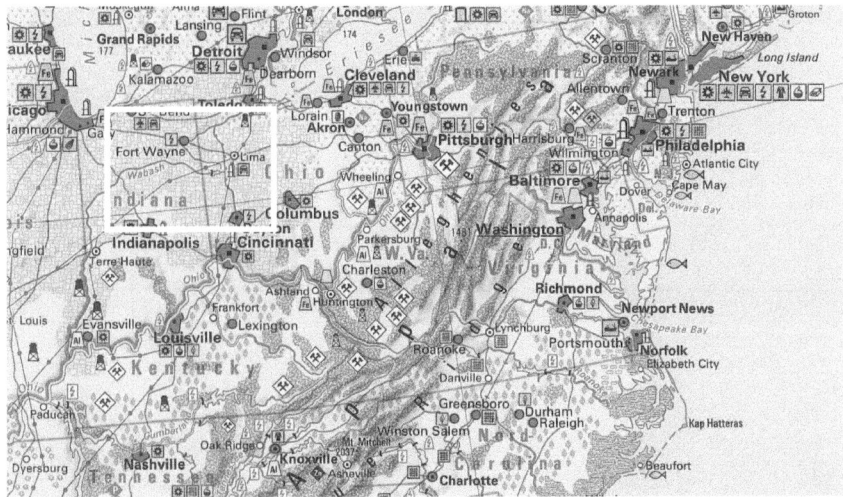

Figure 9.2.21
Land-use map for the reference area grassland / agriculture. Excerpt from Alexander Weltatlas (Schulze, 1989)

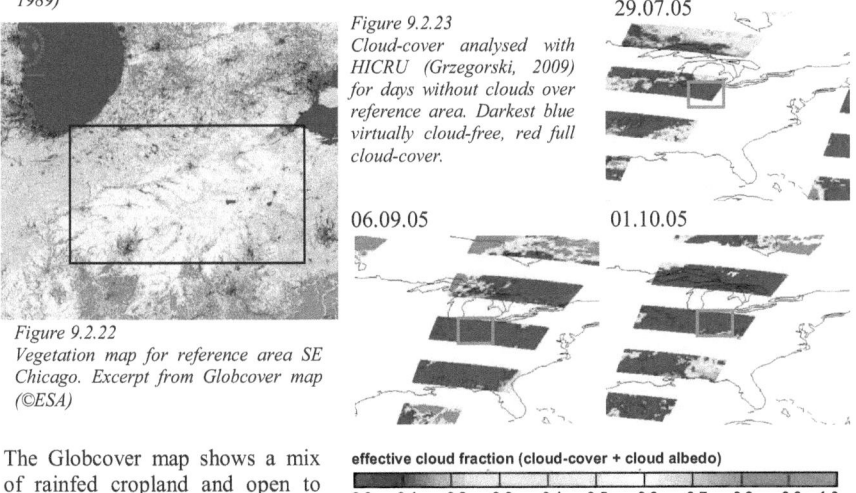

Figure 9.2.22
Vegetation map for reference area SE Chicago. Excerpt from Globcover map (©ESA)

Figure 9.2.23
Cloud-cover analysed with HICRU (Grzegorski, 2009) for days without clouds over reference area. Darkest blue virtually cloud-free, red full cloud-cover.

The Globcover map shows a mix of rainfed cropland and open to closed grasslands. Some small spots of broadleaf tree areas are also visible. This combination suggests an agricultural mix of cereal and fruit production with animal-grazing (meat, milk, leather) and hay production.

The land-use map suggests dominating maize and soy-bean production. Maize is sweetgrass. Soy-beans have not been measured for this thesis. They would probably show as deciduous signal.

Grasslands for animal-grazing and hay-production consist of sweet-grasses. Many of these grasses have a life-cycle of more than one year. In these areas all year round a sweet-grass signal of variable strength should be seen.
Agricultural areas often include some fruit-trees, most of those belonging to Rosaceae.

A general characteristic of agricultural areas - especially when cereals (sweet-grasses) are cultivated - is the strong seasonal change of vegetation cover. It alternates from bare soil and intensive grass-cover to dying vegetation and bare soil again within 5 to 7 months, depending on crop and region. In some cases the growing season might be up to 10 months for plants where seeding starts in autumn.

9.2.1.6 Grassland: unmanaged

The reference area for unmanaged, arid grassland is situated in Botswana.

Figure 9.2.24
Land-use map for the reference area grassland / unmanaged.
Excerpt from Alexander Weltatlas (Schulze, 1989)

This reference area spans the grasslands between the Okavango-Delta and the Kalahari desert. These grasslands are mostly unmanaged. Growth conditions get progressively harder from north to south.

The Globcover map classifies the whole area uniformly as "closed to open grassland > 15% cover". The signal should be mostly sweet-grass. Since these areas are not managed, fresh green grass should dominate the signal for only a relatively short time after the rainy season. The rest of the time that signal should be considerably shielded by long dry grass-leaves.

In contrast to agricultural areas the grass is not cut and ploughed under after harvest, so the amount of pure soil should not alternate by season so strongly. However, since the Globcover only claims more than 15% land-cover, intermittent pure soil may be visible at all times.

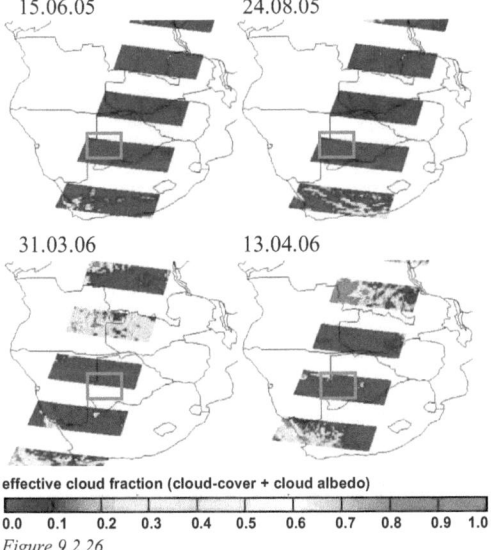

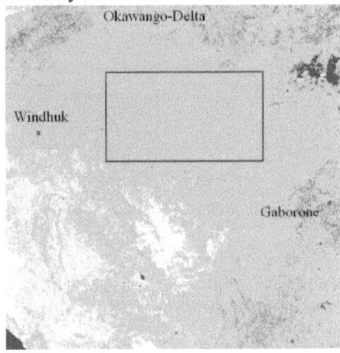

Figure 9.2.26
Cloud-cover analysed with HICRU (Grzegorski, 2009) for days without clouds over reference area. Darkest blue virtually cloud-free, red full cloud-cover.

Figure 9.2.25
Vegetation map for reference area Botswana. Excerpt from Globcover map (©ESA)

9.3 Fit results for the selected reference areas

In this section, the satellite observations were analysed with respect to signatures of vegetation. For that purpose vegetation reference spectra were included in the spectral fitting process as described in section 9.1. Of course, there is some ambiguity with respect to the specific selection of reference spectra, which are considered in the spectral fitting process. As a first choice we used the spectra of Betulaceae, Pinus and Larix.

The respective results are shown in Fig. 9.3.1. Enhanced fit coefficients for the vegetation reference spectra are found for all of the selected areas indicating the sensitivity of our retrieval to the reflectance signals of vegetation. The results for the boreal forest near Hudson Bay are as expected. However, the detailed investigation of the results also shows some ambiguities and inconsistencies: for example, over the rain forest, the signatures of needle trees are strong, while the signature of the one deciduous tree class (Betulaceae) is close to zero. Also, over farmland, mainly enhanced values for the larix spectra were found, which is not expected there.

Other combinations of reference spectra and other reference areas were tested, but overall similar results were found: often not the vegetation signal, which was most expected according to the vegetation classes was found, but other signals showed the highest fit coefficients. Moreover, the results for a specific vegetation reference spectrum were strongly dependent on the specific choice of spectra, which were included in the spectral analysis. These findings indicate the limited information content of the vegetation reference signals. In addition, the ambiguities are probably also a result of the similarity of the spectral structures of different vegetation reference spectra. Finally, as shown in section 8.7 the spectral signatures also strongly depend on the specific viewing geometry and orientation of the leaf surfaces, which might cause deviations of the observed vegetation signals from the vegetation reference spectra measured in this thesis. These findings probably indicate a fundamental limitation of the application of vegetation reference spectra to DOAS satellite remote sensing. In spite of these complications, the results shown in Fig. 9.3.1 indicate that different vegetation signals were observed over the different reference areas. This finding indicates that information on different vegetation signals is indeed contained in the measured satellite spectra.

In the following section we introduce a method, which exploits this information content in a systematic way.

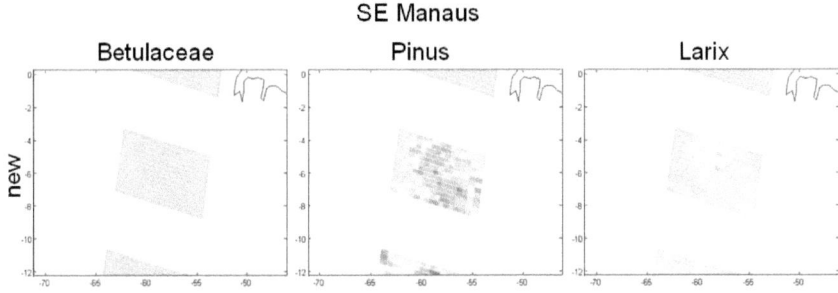
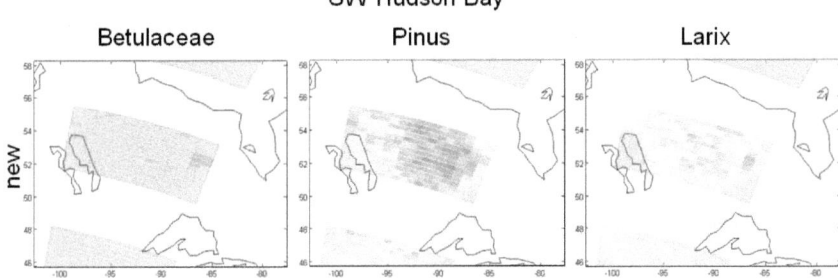
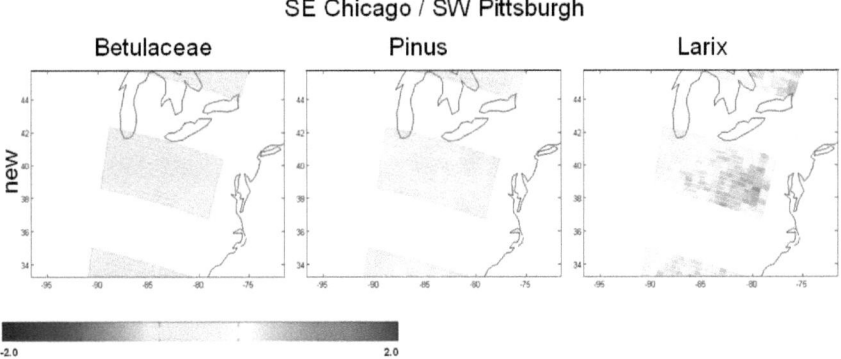

Figure 9.3.1
Results (Fit coefficient of the logarithm of the vegetation reference spectra) of the spectral retrieval for some of the selected reference areas.

9.4 Global monthly mean maps

In this section fit results are analysed for a whole month on a global basis. The month of July was chosen for that purpose, because most continents are located at the northern hemisphere, where the strongest biological activity is found in (northern) summer.

We analysed all measurements, but applied different selection criteria before the results were averaged to yield a global map. First, observations over oceans were excluded. In addition, only measurements with a significant vegetation signal were considered for the monthly mean map. For the determination of the overall vegetation signal, we chose a simplified version of the NDVI, which is especially sensitive to the detection of any vegetation signal. Our vegetation index (vi) is defined as the ratio of radiances at two different wavelengths:

$$(9.1) \quad vi = \frac{R_{715nm}}{R_{680nm}}$$

We only selected measurements with vi < 1. By this choice, not only measurements over cloudy scenes, but also measurements over desert areas were excluded.

In the spectral analysis we included four vegetation reference spectra: Rosaceae, Betulaceae, Larch, Pinus, and Sweet grass. The results for the different vegetation reference spectra are shown in Figures 9.4.1 – 9.4.5.

Vegetation signatures are found for many regions on earth; especially for Betulaceae, Pinus and Larch, enhanced fit coefficients are observed over regions with strong vegetation. Also for Sweet grass, enhanced signals are found, but especially in tropical regions. While these results are in general encouraging, again, no clear distinction between the results for the different vegetation reference spectra are found. Especially for Rosaceae, Pinus and Larch, enhanced values are typically found for the same regions. Also, for Betulaceae only negative fit coefficients were found. Like in the examples shown in the previous section, these results indicate that the used reference spectra do not perfectly match the vegetation signals found in the satellite spectra. Nevertheless, it is also obvious that for specific regions, the fit coefficients for the different vegetation spectra show significant differences. This finding indicates that the satellite spectra in fact contain specific information on different kinds of vegetation (but these signatures are not unambiguously assigned to our vegetation reference spectra). In the following section a method is introduced to identify different vegetation classes from the satellite measurements using the vegetation reference spectra measured in this thesis.

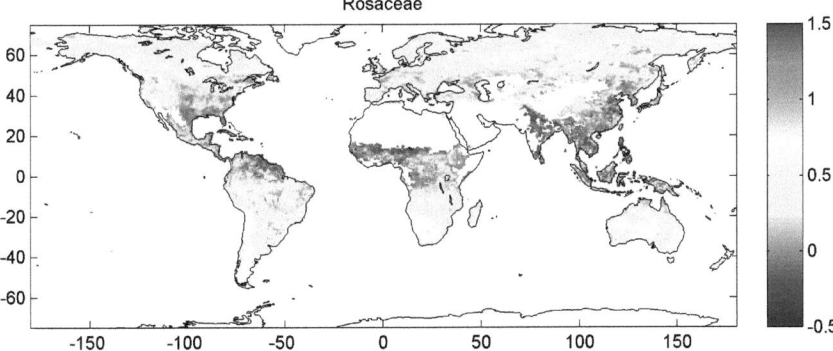

Figure 9.4.1
Monthly means for July 2005 SCIAMACHY-data fitted with Rosaceae. The result is very encouraging. No response in known desert areas, low response in the African Sahel, the monsoon-forest areas of south-east Asia, and most rainforests with the exception of the Amazon-basin. Middle to high response in the good agricultural areas of non-tropical regions and boreal forests.

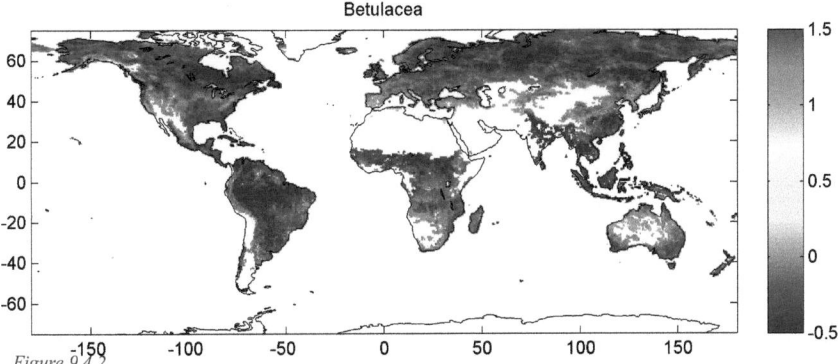

Figure 9.4.2
Monthly means for July 2005 SCIAMACHY-data fitted with Betulaceae. The response is mostly negative. The spectra is used by the fit to compensate not fitted effects. Response suffers from spectral similarity to Rosaceae.

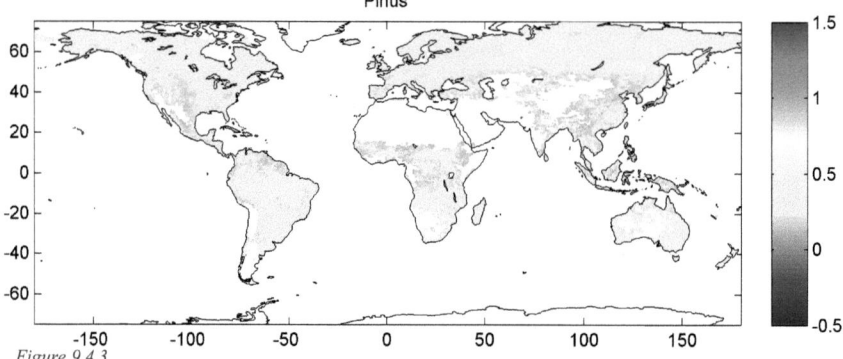

Figure 9.4.3
Monthly means for July 2005 SCIAMACHY-data fitted with Pinus .Middle strong signal in almost all vegetation-zones. Structures very similar to Larch, just stronger. Strongest signals in the boreal forests of Eurasia and North-America. Another maximum can be found in the forests of Brazil.

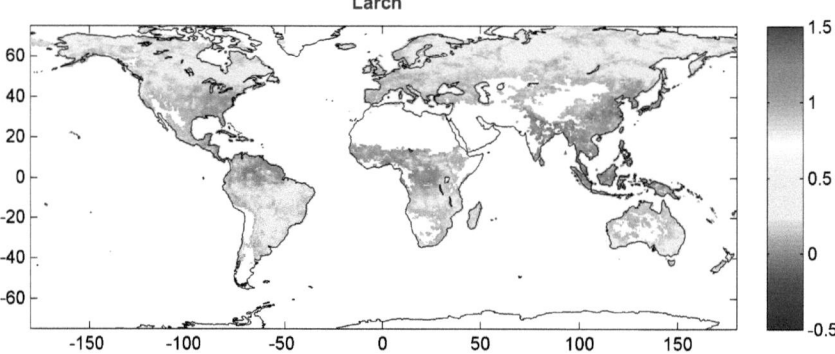

Figure 9.4.4
Monthly means for July 2005 SCIAMACHY-data fitted with Larch. Low to middle signal in almost all vegetation-zones. Almost ubiquitous signal not consistent with Larch. Highest responses over Siberia and other boreal regions is reasonable, though.

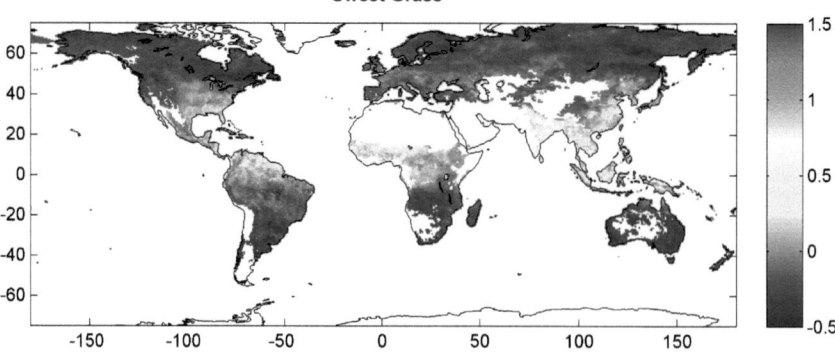

Figure 9.4.5
Monthly means for July 2005 SCIAMACHY-data fitted Sweet Grass. The grass-response is generally low, but strongest in Indonesia and northern tropical margins and subtropical areas. The surprisingly low values in western and central Europe may indicate that many grasses are near the end of their lifecycle, close to harvest. (Compare also chapter 8.4.)

9.5 Modified global monthly mean maps

In this section an improved interpretation of the satellite vegetation retrieval is introduced. It is based on the fit results shown in the previous section, but uses specific linear combinations of the results of the different vegetation reference spectra. This method was developed in cooperation with Steffen Beirle.

The basic assumptions of the new method are:
a) specific information on different vegetation classes is contained in the measured satellite spectra. This was confirmed by the results presented in sections 9. 3 and 9.4.
b) part of the vegetation signal in the satellite spectra is directly assigned to specific vegetation reference spectra (e.g. rosaceae or sweet grass), but there seems to be some ambiguity, how the vegetation signal is assigned to the different reference spectra (from the spectral analysis).
c) Deviations between the vegetation signals in the satellite spectra and the measured reference spectra is related to different reasons, in particular to the dependence on the viewing geometry.

Based on these assumptions, linear combinations of the results for the different vegetation reference spectra were defined. These were determined by forcing the new linear combination of fit results over certain regions to represent specific vegetation-classes. Over the selected areas, the respective linear combination was forced to be 1, while all other classes were forced to 0.

The following reference areas used for these new classifications were chosen as follows:
-rainforests in the Amazon basin, from 0 to -10° S and 55 to 75° W.
-grass in the North-American prairies, from 40 to 45° N and 100 to 105° W.
-coniferous forest, from south-west of Hudson Bay at 51 to 55° N and 90 to 95° W.
-agriculture, a small region south-east of Chicago, from 40 to 41° N and 83 to 87° W.
The linear combinations, which best fulfilled the above criteria are summarised in Tab. 9.4.

Coefficients of the linear fit:

	Larch	Sweet Grass	Rosaceae	Pinus	Betulaceae	global offset
Rainforest	-5.2009	-0.3602	-0.4387	0.1970	3.3639	0.2980
Grass	9.2023	8.8697	5.5823	-13.9424	0.6936	2.1770
Conifers	5.0968	0.5387	-2.7275	7.6135	1.6189	-1.6809
Agriculture	1.3587	2.4709	-0.6682	9.1344	4.8242	-1.1284

Table 9.4
Linear fit coefficients for four vegetation-classes.

The coefficients for the class Grass indicates that the reference area was poorly chosen. Indeed the area includes a forested mountain range. These ambiguities have to be considered when interpreting the results for the Grass-class.

The respective results of the monthly averaged vegetation signals for July and August 2005 are shown in Figure 9.4.6.

The new method largely improved the separation between the different (new) vegetation classes. The four selected vegetation classes show distinct spatial patterns in good agreement with the expected distributions. Most importantly, consistent signals are not only found for the regions, for which the linear combinations were forced, but also for regions at distant locations. This indicates that different vegetation classes can indeed be identified by their spectral signatures in the red spectral range. Further refinements of the method can be expected if larger data sets are used for the determination of the linear combinations (including also different seasons).

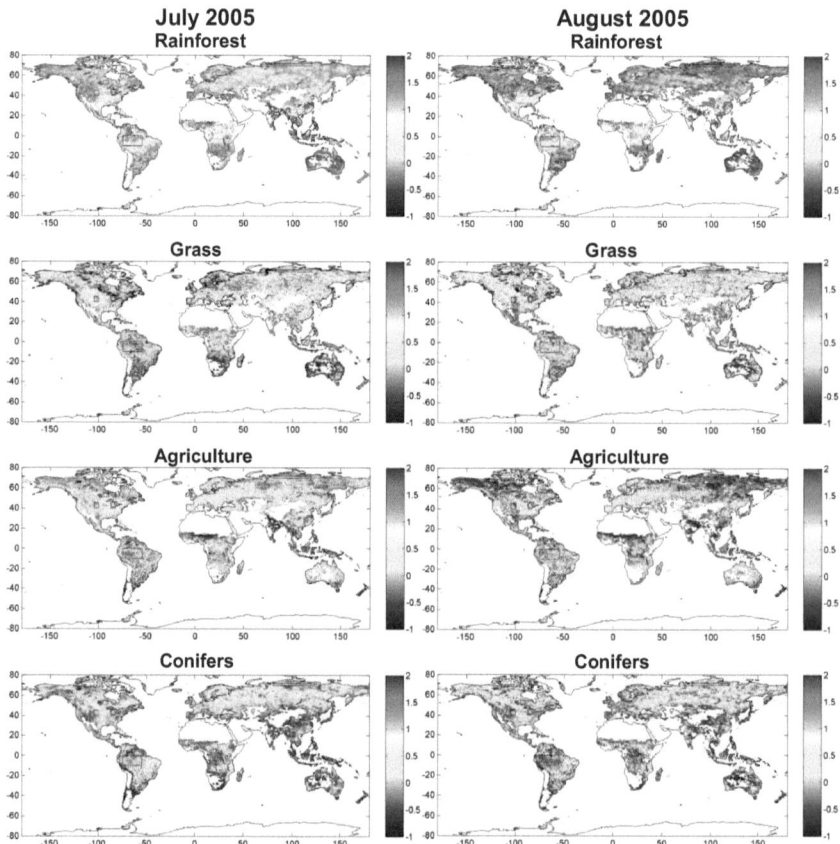

Figure 9.4.6 Classification results for July and August 2005 for the vegetation-classes Rainforest, Grass, Agriculture and Conifers. Source: Beirle (personal communication)

10 Conclusion and Outlook

The goal of this thesis has been to investigate the potential of Differential Optical Absorption Spectroscopy (DOAS) for monitoring global vegetation. The first results using previously available reference spectra were encouraging, but new and more suitable, vegetation reference spectra had then to be created.

Particular care was taken to exclude any conceivable external influence on the reference spectral measurements. For example, an artificial light source was used to measure the reflection spectra of different kinds of vegetation. In this thesis, reflection spectra for a large number of plant species were derived and systematically analysed. The set of reference spectra obtained is unique in its extent and also with respect to its spectral resolution and the quality of the spectral calibration. For the first time, this has allowed a comprehensive investigation of the high-frequency spectral structures of vegetation reflectance and of their dependence on the viewing geometry (see also Chapter 8.)

The measurements show the highest variability in the wavelength-range of 630 to 730 nm, in particular in the major chlorophyll absorption range of 660 to 700 nm. The spectral structures in the wavelength-range above 730 nm are particularly sensitive to changes in illumination and viewing-angle.

The results indicate that high-frequency reflectance from vegetation is very complex and highly variable. While this is an interesting finding in itself, it also complicates the application of the obtained reference spectra to the spectral analysis of satellite observations. In particular, none of the cases examined that are studied in Chapter 8 will be the sole case within any given satellite ground pixel. Moreover, leaves exhibit different orientations with respect to the illumination of the sun and the view of the satellite instrument, because of natural growing conditions and – depending e.g. on the wind – some leaves get turned while others remain in place. Plants have different periods of maximum growth – within species according to micro-climate, water supply and drainage, soil-quality etc. and also between species – and the onset of autumn-colouring (if at all). Some of these effects may cancel each other out, while others might lead to systematic deviations of the vegetation signals measured from a satellite compared to those of the reference data sets.

Indeed, including the generated reference spectra in the spectral analysis of satellite measurements leads to partly ambiguous results. While, in general, the presence and amount of vegetation was well recognised by the enhanced fit coefficients for the different vegetation reference spectra, the results did not always represent the expected vegetation classes (see Chapter 9). This indicates that possible deviations of the reference spectra from the signatures in the satellite spectra (e.g. caused by variation of the leaf orientation) are partly compensated by reflectance signatures in other vegetation reference spectra. On the other hand, the results obtained also indicated that the satellite spectra do contain spectral information on different kinds of vegetation, because different vegetation signals were found at different locations on earth.

One way to deal with this ambiguity was to build new linear combinations of the vegetation spectra used by forcing the fit results to the expected values over selected regions. This method was successfully applied for four selected regions for two selected months (July and August 2005) with high biological activity in the northern hemisphere. The vegetation signals of the reference areas could be well identified as well as those for other regions, and consistent patterns were identified over the whole earth. This method could be further improved by using refined vegetation spectra (see Chapter 6) and by choosing multiple - but small and carefully selected - training areas for different vegetation-classes.

Another aspect, which should be further investigated, is that vegetation develops in cycles. For example, the response of grasses changes significantly with season and deciduous trees alternate between stages of leaf-development and bare branches, with a colourful transition phase in autumn.

The new set of vegetation reference spectra created in this thesis opens new perspectives for research. Besides refined satellite analyses, these spectra might also be used for applications on other platforms such as aircraft. First promising studies have been presented in this thesis, but the full potential for the remote sensing of vegetation from satellite (or aircraft) could be further exploited in future studies.

References

Anderson, J. M. (1986) Photoregulation of the composition, function, and structure of thylakoid membranes. Ann. Rev. Plant Physiol., 37, 93–136.

Anderson, J. M., Chow, W. S. & Park, Y. I. (1995) The grand design of photosynthesis: acclimation of the photosynthetic apparatus to environmental cues. Photosynth. Res., 46, 129–139.

Arino, O., Leroy, M., Ranera, F., Gross, D., Bicheron, P., Nino, F., Brockman, C., Defourny, P., Vancutsem, C., Achard, F., Durieux, L., Bourg, L., Latham, J., Di Gregori, A., Witt, R., Herold, M., Sambale, J., Plummer, S., Weber, J-L., Goryl, P., Houghton, N. : Globcover - A Global Land Cover Service with MERIS. http://www.gofc-gold.uni-jena.de/documents/boston/Arino_Globcover_ENVISATSYMPOSIUM.pdf (19.10.10)

ASTER spectral library: http://speclib.jpl.nasa.gov/

AT2-ELS: Authors: Annette Ladstätter-Weißenmayer (IUP, University of Bremen), Andreas Richter (IUP, University of Bremen), Maria Kanakidou (ECPL, University of Crete). With additional input from Thomas Wagner (University of Heidelberg), John P. Burrows (IUP, University of Bremen) and the AT2 group:
E-Learning Radiative Transfer Equation (23.01.2008)
E-Learning Radiative Transfer in the Atmosphere (11.03.2009)
E-Learning Retrieval procedures (11.03.2009)
http://dev1.nilu.no/moodle/at2/exerciseTutor/ExerciseModule.htm?/moodle/at2-els_NO2/at2-els_NO2.htm (23.01.2008 and 11.03.2009)
http://www.iup.uni-bremen.de/E-Learning/at2-els_NO2/index.htm (13.10.2010)

Barrett, EC., Curtis, L.F. (1992): Introduction to Environmental Remote Sensing. Third edition. London,UK

Bicheron, P., Defourny, P., Brockmann, C., Schouten, L., Vancutsem, C., Huc, M., Bontemps, S., Leroy, M., Achard, F., Herold, M., Ranera, F., Arino, O. (2008) GLOBCOVER. Products Description and Validation Report. http://ionia1.esrin.esa.int/docs/GLOBCOVER_Products_Description_Validation_Report_I2.1.pdf

Böhlmann, D. (2009): Warum Bäume nicht in den Himmel wachsen. Eine Einführung in das Leben unserer Gehölze. Wiebelsheim, Germany.

Bohren, C.F. (2001): Clouds in a Glass of Beer. Simple Experiments in Atmospheric Physics. Mineola, NY, USA.

Bouvier, F., Suire, C., Mutterer, J. & Camara, B. (2003) Oxidative remodeling of chromoplast carotenoids: identification of the carotenoid dioxygenase CsCCD and CsZCD genes involved in crocus secondary metabolite biogenesis. Plant Cell, 15, 47–62.

Boyle, R. (1664) Experiments and Considerations Touching Colours: First Occasionally Written Among Some Other Essays to a Friend and now Suffer'd to Come Abroad as the

Beginning of an Experimental History of Colours. Henry Herringman, London, UK. Available on microfilm at Early English Books Online, http://www.lib.umich.edu/eebo/

Bracher, A., Vountas, M., Dinter, T., Burrows, J.P., Röttgers, R., Peeken, I. (2009) Quantitative observation of cyanobacteria and diatoms from space using PhytoDOAS on SCIAMACHY data, Biogeosciences, 6, 751-764.

Bramley, P. M. (2002) Regulation of carotenoid formation during tomato fruit ripening and development. J. Exp. Bot., 53, 2107–2113.

Bremer, H. (1999): Die Tropen. Geographische Synthese einer fremden Welt im Umbruch. Berlin, Germany.

Britton, G. (1995) UV/Visible Spectroscopy, in Carotenoids, Vol. 1B: Spectroscopy (eds G. Britton, S. Liaaen-Jensen & H. Pfander), Birkhäuser, Basel, Switzerland, pp. 13–62.

Brydegaard, M., Guan, Z., Wellenreuther, M.; and Svanberg, S: (2009) Insect monitoring with fluorescence lidar techniques: feasibility study. Appl. Opt. **48**, 5668-5677 (2009) http://www.opticsinfobase.org/abstract.cfm?URI=ao-48-30-5668

Carter, G.A. (1991) Primary and Secondary Effects of the Water content on the Spectral Reflectance of Leaves. American Journal of Botany, 78(7); pp. 916-924.

Christopherson, R.W. (1994) Geosystems. An Introduction to Physical Geography. Second Edition. New York, USA.

Curran, P.J. (1992) Principles of Remote Sensing. New York, USA.

Curtis, L.F. (1978): Remote sensing systems for monitoring crops and vegetation. Progress in Physical Geography 1978; 2; 55. DOI: 10.1177/030913337800200104

Davies, K.M. (Editor) (2004): Plant Pigments and their Manipulation. Oxford, UK.

Eigemeier, E., Beirle, S., Platt, U., Schubert, A., Wagner, T. (2008)Global monitoring of annual vegetation cycles using DOAS satellite observations. Proceedings from 2nd MERIS/(A)ATSR Workshop, Frascati, Italy. ESA Special Publication SP-666, (2008)

Eigemeier, E. , Beirle, S., Platt,U., Preusser, A., Wagner,T. (2009) Monitoring vegetation using DOAS satellite observations: creating a set of reference spectra. Proceedings EUMETSAT Conference in Bath, UK.

Eigemeier, E. , Beirle, S., Marbach, T., Platt,U., Wagner,T. (2010) Monitoring vegetation using DOAS satellite observations based on new reference data. Proceedings ESA Living Planet Symposium in Bergen, Norway.

Formaggio, E., Cinque, G. & Bassi, R. (2001) Functional architecture of the major light-harvesting complex from higher plants. J. Mol. Biol., 314, 1157–1166.

Gausman, H.W. (1977): Reflectance of Leaf Components. Remote Sensing of Environment 6, 1-9 (1977)

Gomer, T., Brauers, T. Heintz, F., Stutz, J. & Platt, U. (1993) MFC user manual, version 1.98. inhouse publication, Institut für Umweltphysik, University of Heidelberg, Germany.

Gould, K. S., Vogelmann, T. C., Han, T. & Clearwater, M. J. (2002a) Profiles of photosynthesis within red and green leaves of Quintinia serrata A. Cunn. Physiol. Plant., 116, 127–133.

Gross, J. (1991) Pigments in Vegetables, Van Nostrand Reinhold, New York, USA.

Grzegorski, M (2009) Cloud retrieval from UV/vis satellite instrument. PhD thesis, University of Heidelberg, Germany

Hall, D.O., Rao, K.K. (1999): Photosynthesis. Sixth edition. Cambridge, UK.

Harborne J. B. (1988) The flavonoids: recent advances, in Plant Pigments (ed. T. W. Goodwin), Academic Press, London, UK, pp. 299–343.

Herrin, D. L., Battey, J. F., Greer, K. & Schmidt, G. W. (1992) Regulation of chlorophyll apoprotein expression and accumlation. Requirements for carotenoids and chlorophyll. J. Biol. Chem., 267, 8260–8269.

Hess, D. (1991) Pflanzenphysiologie. UTB, 9. durchgesehene Auflage. Stuttgart, Germany.

HITRAN database: http://www.cfa.harvard.edu/HITRAN/

Hoch, W. A., Zeldin, E. L. & McCown, B. H. (2001) Physiological significance of anthocyanins during autumnal leaf senescence. Tree Physiol., 21, 1–8.

Hofmeister, H. (1987): Lebensraum Wald. 2., revidierte Aufl. Hamburg, Germany. ISBN 3-490-16818-6

http://de.wikipedia.org/wiki/Chlorophyll (07.10.2010)

http://joseba.mpch-mainz.mpg.de/pdf_dateien/ground_based_UVvis.pdf (13.10.2010)

http://joseba.mpch-mainz.mpg.de/pdf_dateien/uv_vis_sat_1.pdf (13.10.2010)

http://joseba.mpch-mainz.mpg.de/pdf_dateien/interaction.pdf (13.10.2010)

Hulst, V.C. van de (1981) Light scattering by small particles. New York, NY, USA.

Jensen, J.R. (2000) Remote Sensing of the Environment. An Earth Resource Perspective. Upper Saddle River, NJ, USA.

Jordan, B. R. (1996) The effects of ultraviolet-B radiation on plants: a molecular perspective, in Advances in Botanical Research, Vol. 22 (ed. J. A. Callow), Academic Press, London, UK, pp. 97–162.

Kalensky, Z., Wilson, D.A. (1975) Spectral Signature of Forest Trees. Proceedins: Third Canadian Symposium on Remote Sensing, pp. 155-171.

Kautsky, H., Hirsch, A. (1931), Neue Versuche zur Kohlensäureassimilation. Naturwissenschaften 19: 964

Larcher, W. (1984) Ökologie der Pflanzen auf physiologischer Grundlage. UTB. 4. Auflage, Stuttgart, Germany.

Lee, D. W. (2002) Anthocyanins in autumn leaf senescence. Adv. Bot. Res., 37, 147–165.

Leue, C. (1999) Globale Bilanzierung der NOX-Emissionen aus GOME Satelliten-Bildfolgen, PhD-thesis, University of Heidelberg, Germany.

Kay, Q. O. N., Daoud, H. S. & Stirton, C. H. (1981) Pigment distribution, light reflection and cell structure in petals. Bot. J. Linnean Soc., 83, 57–84.

Klink, H.-J. (1996): Vegetationsgeographie. Das Geographische Seminar. 2. Auflage.

Knipling, E.B. (1970): Physical and Physiological Basis for the Reflectance of Visible and Near-Infrared Radiation from Vegetation. Remote Sensing of Environment 1 (1970), 155-159

Lillesand, T.M. & Kiefer, R.W. (1987) Remote Sensing and Image Interpretation. Second Edition. New York, USA.

Markham, K. R., Gould, K. S., Winefield, C. S., Mitchell, K. A., Bloor, S. J.&Boase, M. R. (2000) Anthocyanic vacuolar inclusions – their nature and significance in flower colouration. Phytochemistry, 55, 327–336.

Matile, P., Hortensteiner, S. & Thomas, H. (1999) Chlorophyll degradation. Ann. Rev. Plant Physiol. Plant Mol. Biol., 50, 67–95.

Matile, P. (2000) Biochemistry of an Indian summer: physiology of autumnal leaf coloration. Exp. Gerontology, 35, 145–158.

Matsui, S., Suzuki, T. & Nakamura, M. (1984) Distribution of flower pigments in perianth of Vandeae orchids. Res. Bull. Fac. Ag. Gifu Uni., 49, 361–369.

Matsui, S. & Nakamura, M. (1988) Distribution of flower pigments in perianth of Cattleya and allied genera. I. Species. J. Jap. Soc. Hort. Sci., 57, 222–232.

Nachtigall, W. (1985) Unbekannte Umwelt. München, Germany.

Nultsch, W. (1986): Allgemeine Botanik. 8. Auflage. Stuttgart, Germany

Ocean Optics (2007): USB2000+ Fiber Optic Spectrometer. Installation and Operation Manual. Document Number 270-00000-000-02-0807.

Parker, A.R. & Martini, N. (2005): Structural colour in animals—simple to complex optics. Optics & Laser Technology 38 (2006) 315–322. doi:10.1016/j.optlastec.2005.06.037

Pecket, R. C. & Small, C. J. (1980) Occurrence, location and development of anthocyanoplasts. Phytochemistry, 19, 2571–2576.

Platt, U. (1994) Differential Optical Absorption Spectroscopy (DOAS), in Air monitoring by spectrometric techniques, edited by M. Sigrist, Chemical Analysis Series, John Wiley, New York (1994), 127, 27 – 84

Platt, U. and Stutz, J. (2008) Differential Optical Absorption Spectroscopy: Principles and Applications, Heidelberg, Germany

Porra, R. J. (2002) The checkered history of the development and use of simultaneous equations for the accurate determination of chlorophylls a and b. Photosynth. Res., 73, 149–156.

Rast, M. (1991) Imaging Spectroscopy and its Application in Spaceborne Systems. ESA SP-1144, ISBN 92-9092-152-8, December 1991

Rautiainen, M. (2005) The spectral signature of coniferous forests: the role of stand structure and leaf area index. Dissertation at University of Helsinki, Helsinki, Finland.

Rees, W.G. (1990) Topics in Remote Sensing 1. Physical Principles of Remote Sensing. Cambridge, UK.

Reinbothe, S., Reinbothe, C., Apel, K. & Lebedev, N. (1996) Evolution of chlorophyll biosynthesis – the challenge to survive photooxidation. Cell, 86, 703–705.

Richter, G. (1988): Stoffwechselphysiologie der Pflanzen. Physiologie und Biochemie des Primär- und Sekundärstoffwechsels. 5. Auflage. Stuttgart, Germany.

Schoefs, B. (2002) Chlorophyll and carotenoid analysis in food products. Properties of the pigments and methods of analysis. Trends Food Sci. Tech., 13, 361–71.

Schulze, H. (Hrsg.) (1989): Alexander Weltatlas. Neue Gesmatausgabe. Stuttgart, Germany.

Seinfeld, J.H. & Pandis, S.N. (2006) Atmospheric Chemistry and Physics. From Air Pollution to Climate Change. 2nd Edition. Hoboken, NJ, USA.

Slater, P.N. (1980) Remote Sensing: Optics and Optical Systems. Reading, Pennsylvania, USA.

Solovchenko, A. & Schmitz-Eiberger, M. (2003) Significance of skin flavonoids for UV-B-protection in apple fruits. J. Exp. Bot., 54, 1977–1984.

Steyn, W. J., Wand, S. J. E., Holcroft, D. M. & Jacobs, G. (2002) Anthocyanins in vegetative tissues: a proposed unified function in photoprotection. New Phytologist, 155, 349–361.

Strack, D. & Wray, V. (1989) Anthocyanins, in Methods in Plant Biochemistry (eds P. M. Dey & J. B. Harborne), Academic Press, London, UK, pp. 325–355.

Strasburger, E. (1991): Lehrbuch der Botanik. 33. Auflage, Stuttgart, Germany.

Stutz, J. (1996) Messung der Konzentration troposphärischer Spurenstoffe mittels Differentieller-Optischer-Absorptionsspektroskopie: Eine neue Generation von Geräten und Algorithmen. PhD-thesis, Ruprecht Karls Universität Heidelberg, Germany.

Swain, T. (1976) Nature and properties of flavonoids, in Chemistry and Biochemistry of Plant Pigments, Vol.1 (ed. T.W. Goodwin), Academic Press, New York, pp. 425–463.

Vetter, W., Englert, G., Rigassi, N. & Schwieter, U. (1971) Spectroscopic methods, in Carotenoids (ed. O.Isler), Birkhaüser, Basel, Switzerland, pp. 189–266.

von Lintig, J., Welsch, R., Bonk, M., Giuliano, G., Batschauer, A. & Kleinig, H. (1997) Light-dependent regulation of carotenoid biosynthesis occurs at the level of phytoene synthase expression and is mediated by phytochrome in Sinapis alba and Arabidopsis thaliana seedlings. Plant J., 12, 625–634.

von Wettstein, D. (2000) Chlorophyll biosynthesis I: from analysis of mutants to genetic engineering of the pathway. Discoveries Plant Biol., 3, 75–93.

Vountas, M., Dinter, T., Bracher, A., Burrows, J. P. and Sierk, B. (2007) Spectral studies of ocean water with space-borne sensor SCIAMACHY using Differential Optical Absorption Spectroscopy (DOAS). Ocean Sci., 3, 429–440, 2007, www.ocean-sci.net/3/429/2007/

Walter, H., Breckle, S.-W. (1999): Vegetation und Klimazonen. UTB. 7. Auflage. Stuttgart, Germany.

Wagner, T. (1999) Satelliet observations of atmospheric halogen oxides. Ph.D. thesis, University of Heidelberg, Germany. http://www.ub.uni-heidelberg.de/archiv/539

Wagner, T., Beirle,S., Deutschmann,T., Grzegorski, M., and Platt,U. (2007) Satellite monitoring of different vegetation types by differential optical absorption spectroscopy (DOAS) in the red spectral range, Atmos. Chem. Phys., 7, 69-79, (2007)

Wagner, T., Beirle,S., Deutschmann,T., Eigemeier, E., Frankenberg, C., Grzegorski, M., Liu, C., Marbach, T., Platt,U. and Penning de Vries,M. (2008) Monitoring of atmospheric trace gases, clouds, aerosols and surface properties from UV/vis/NIR satellite instruments. J. Opt. A: Pure Appl. Opt. 10 (2008) 104019 (9 pp)

Wagner, T., Beirle, S., Deutschmann, T., Frankenberg, C., Grzegorski, M., Khokhar, M.F., Kühl, S., Marbach, T., Mies, K., Penning de Vries, M., Platt, U., Pukite, J. (2008 b) Spione für atmosphärische SchadstoffeundTreibhausgase:neue Satelliteninstrumente ermöglichen globalen Blick. http://joseba.mpch-mainz.mpg.de/pdf_dateien/twagner_speyer_2008.pdf

Welsch, R., Beyer, P., Hugueney, P., Kleinig, H. & von Lintig, J. (2000) Regulation and activation of phytoene synthase, a key enzyme in carotenoid biosynthesis, during photomorphogenesis. Planta, 211, 846–854.

Wild, A., Rothe, G., Vollenweider, G. & Zerbe, R. (1993) Pflanzenphysiologisches Praktikum. Interne Praktikumsanleitung. Institut für Allgemeine Botanik der Johannes Gutenberg Universität Mainz, Germany.

Willows, R. D. (2003) Biosynthesis of chlorophylls from protoporphyrin IX. Nat. Prod. Rep., 20, 327–241.

Young, A. J. & Frank, H. A. (1996) Energy transfer reactions involving carotenoids: quenching of chlorophyll fluorescence. J. Photochem. Photobiol., B: Biology, 36, 3–15.

Appendix A: Measured Vegetation

This appendix shows the names and dates of the vegetation measurements done for this thesis. Where possible, photographs show the plant and the corresponding identification plate from the botanic garden.

Sweet Grasses

german name	botanic name	botanic family	date measured	english name
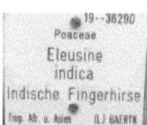	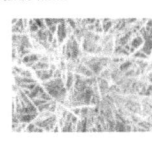		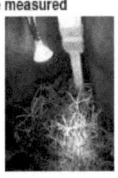	
indische Fingerhirse	Eleusine indica	Poaceae	04.09.09	Indian goosegrass
			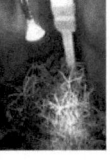	
Perlhirse	Pennisetum glaucum	Poaceae	14.08.09, 21.08.09, 14.09.09, 19.10.09	Pearl millet

Chinaschilf	Miscanthus sinensis	Poaceae	14.09.09	Chinese silver grass

Wilde Mohrenhirse	Sorghum halepense	Gramineae	14.09.09	Johnsongrass

Mohrenhirse	Sorghum durra	Poaceae	14.09.09, 20.10.09	

Mohrenhirse	Sorghum bicolor	Gramineae	14.09.09, 20.10.09, 24.10.09	

german name	botanic name	botanic family	date measured	english name
Besen-Hirse	Sorghum dochna var. technicum	Gramineae	20.10.09	
Japanische Hirse	Echinochloa utilis	Gramineae	15.09.09	
Rispenhirse	Panicum miliaceum	Poaceae	24.10.09	Proso millet, common millet
Mais	Zea mays	Poaceae	14.07.09, 14.09.09	Maize
Pfahlrohr oder Riesenschilf	Arundo donax	Gramineae	14.09.09	Giant cane
Gemeines Schilf	Phragmites australis	Gramineae	14.09.09	Common reed

german name	botanic name	botanic family	date measured	english name
Kolbenhirse	Setaria italica	Gramineae	14.09.09, 15.09.09	Foxtail millet
Flaschenbürstengras	Hystrix patula	Gramineae	15.09.09	
Plattährengras	Chasmanthium latifolium		14.08.09, 21.08.09	Woodoats / River oats
Schlafgras	Stipa robusta	Poaceae	15.09.09	Sleepy grass
Riesiges Federgras	Stipa gigantea	Gramineae	15.09.09	Giant feather grass
Stipa viridula		Poaceae	15.09.09	Green needle grass

german name	botanic name	botanic family	date measured	english name
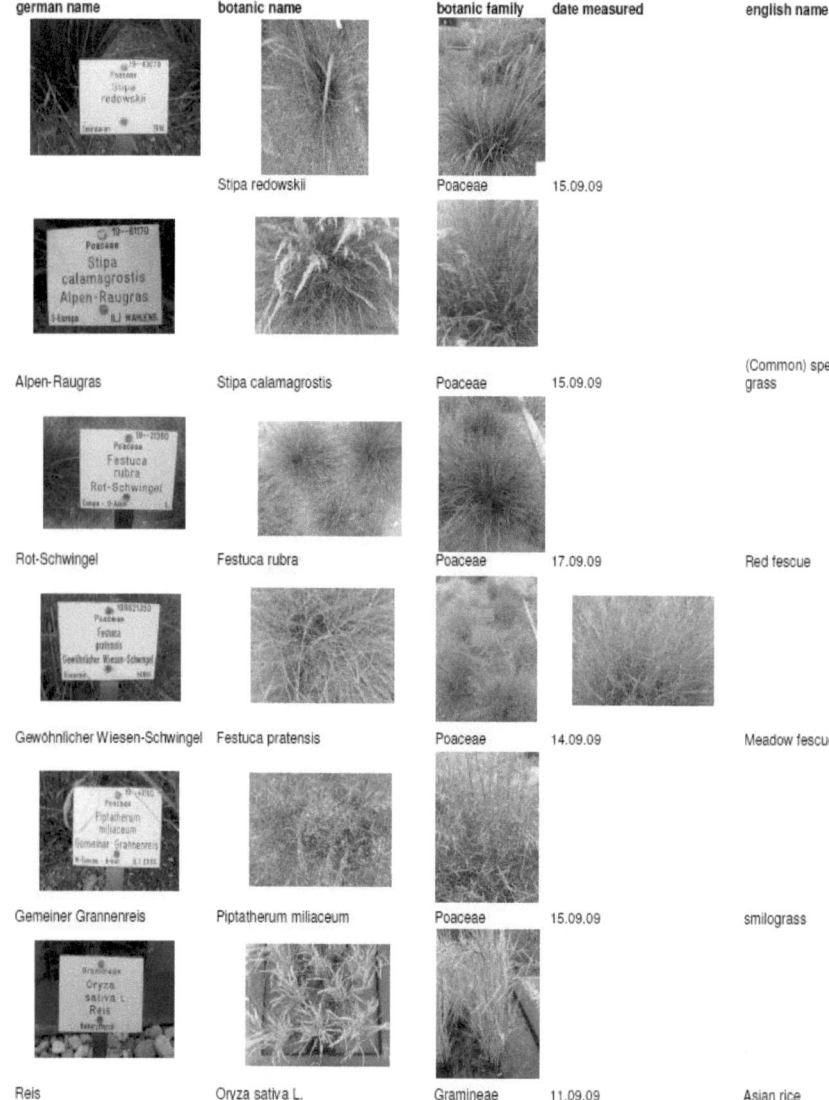				
	Stipa redowskii	Poaceae	15.09.09	
Alpen-Raugras	Stipa calamagrostis	Poaceae	15.09.09	(Common) spear grass
Rot-Schwingel	Festuca rubra	Poaceae	17.09.09	Red fescue
Gewöhnlicher Wiesen-Schwingel	Festuca pratensis	Poaceae	14.09.09	Meadow fescue
Gemeiner Grannenreis	Piptatherum miliaceum	Poaceae	15.09.09	smilograss
Reis	Oryza sativa L.	Gramineae	11.09.09	Asian rice

german name	botanic name	botanic family	date measured	english name
Muriels Schirmbambus	Fargesia murieliae 'Simba'	Poaceae	15.09.09	
Narihirabambus	Semiarundinaria fastuosa	Gramineae	13.11.09	
	Pleioblastus chino	Gramineae	18.11.09	
	Phyllostachys bambusoides 'Castillor	Gramineae	18.11.09	
Knollen-Gerste	Hordeum bulbosum	Poaceae	15.09.09	
Gebirgs-Roggen	Secale montanum	Poaceae	15.09.09	

german name	botanic name	botanic family	date measured	english name
Buntes Perlgras	Melica picta	Gramineae	15.09.09	
Wimper-Perlgras	Melica ciliata	Poaceae	15.09.09	hairy melic, silky spike melic
Fieder-Zwenke	Brachypodium pinnatum	Gramineae	15.09.09	Tor-grass
Aufrechte Trespe	Bromus erectus	Poaceae	15.09.09	Upright Brome, Erect Brome
Wolliges Honiggras	Holcus lanatus	Poaceae	17.09.09	Yorkshire Fog/ Velvet Grass
Rohr-Glanzgras	Phalaris arundinaceae	Gramineae	17.09.09	Reed canary grass

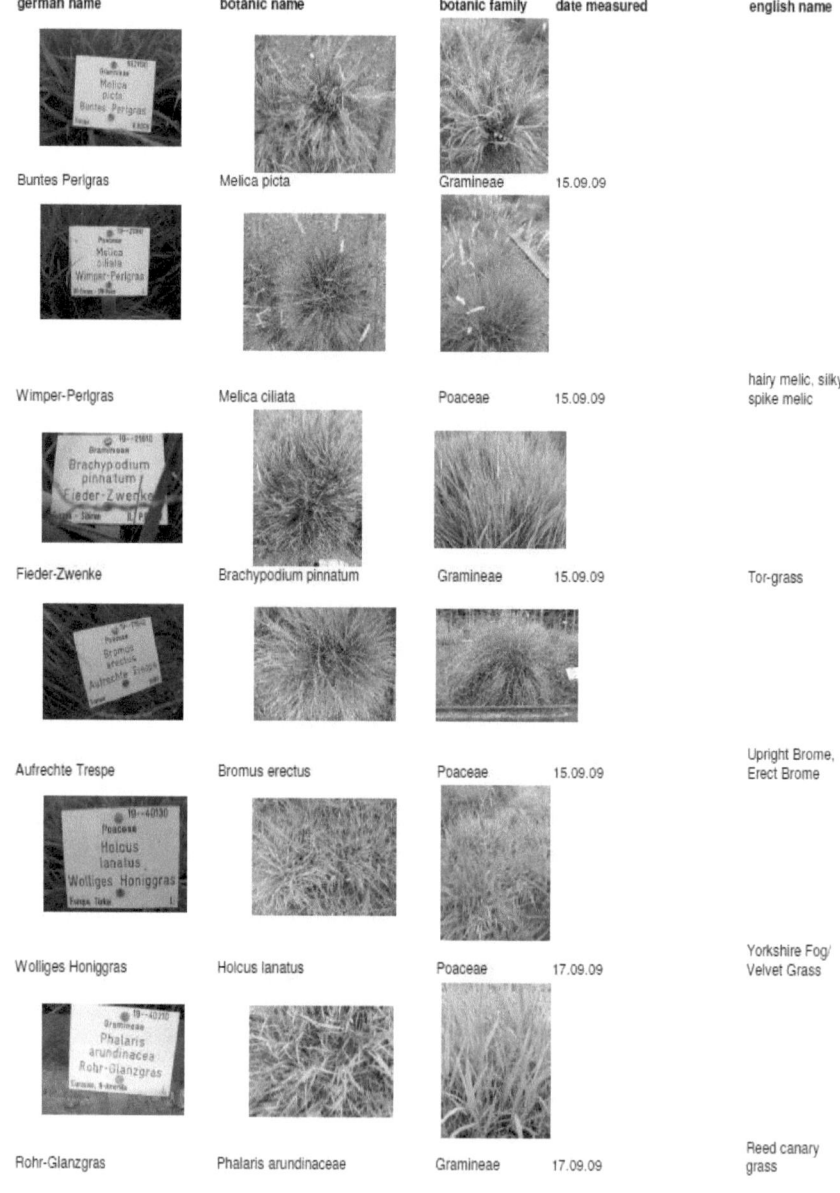

german name	botanic name	botanic family	date measured	english name

Wald Flattergras	Milium effusum	Gramineae	17.09.09	American millettgrass
Land-Reitgras	Calamagrostis epigejos	Poaceae	17.09.09	Wood-small reed
Wiesen-Fuchsschwanzgras	Alopecurus pratensis	Poaceae	17.09.09	Field Meadow Foxtail
Gamagras	Tripsacum dactyloides	Poaceae	19.10.09	Gama Grass
	Pleioblastus gramineus	Gramineae	13.11.09	
	Indocalamus latifolius	Gramineae	18.11.09	

Plants measured 2009

Grasses

german name	botanic name	botanic family	date measured	english name
Schmalblättr. Wollgras	Eriophorum angustifolium	Cyperaceae	15.09.09	Common Cottongrass
Nadel-Sumpfsimse	Eleocharis acicularis	Cyperaceae	09.09.09	Needle Spikerush
Wald-Simse	Scirpus sylvaticus	Cyperaceae	15.09.09	Wood Club-rush
Gewöhnliche Strandsimse	Bolboschoenus maritimus	Cyperaceae	15.09.09	Sea clubrush
Weiße Hainsimse	Luzola luzuloides	Juncaceae	14.09.09	White Wood-rush
Strand-Binse	Juncus maritimus	Juncaceae	16.09.09	Sea rush

german name	botanic name	botanic family	date measured	english name
Baltische Binse	Juncus balticus	Juncaceae	16.09.09	Baltic rush
Platthalm-Binse oder Knollenbinse	Juncus compressus	Juncaceae	16.09.09	Round-fruited rush
Glieder-Binse	Juncus articulatus	Juncaceae	16.09.09	Jointed rush
Flatter-Binse	Juncus effusus	Juncaceae	16.09.09	Common rush/ Soft rush
Salz-Binse	Juncus gerardii	Juncaceae	10.09.09, 16.09.09	Saltmarsh rush
Blaugrüne Binse	Juncus inflexus	Juncaceae	11.09.09, 16.09.09	European meadow rush
Schwarze Kopfbinse	Schoenus nigricans	Cyperaceae	16.09.09	Black bogrush

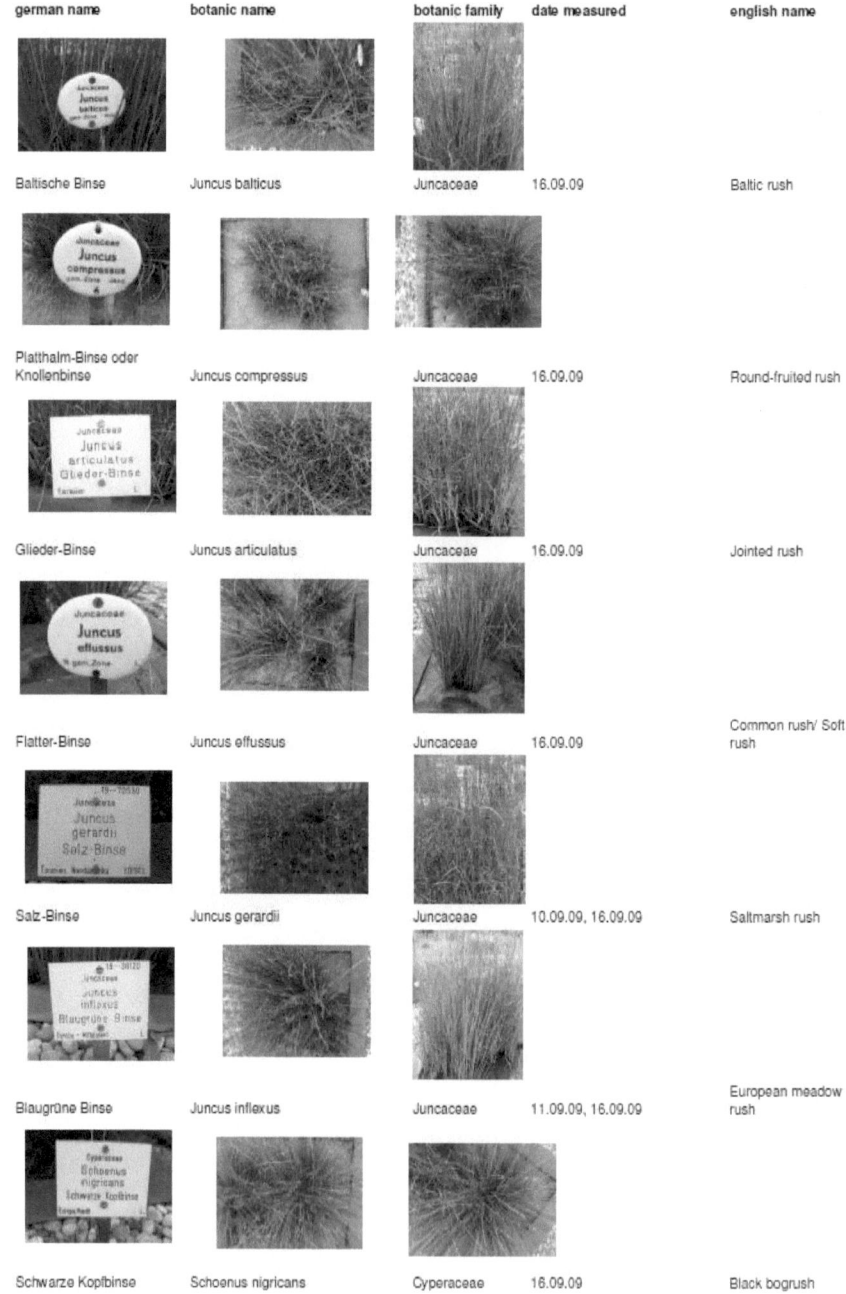

german name	botanic name	botanic family	date measured	english name
Flache Quellbinse	Blysmus compressus	Cyperaceae	07.09.09	Falt sedge
Weiße Segge	Carex alba	Cyperaceae	09.09.09, 15.09.09	
Frühlings-Segge	Carex caryophyllea (La Tourr.)	Cyperaceae	09.09.09, 15.09.09	
Sparrige Segge	Carex muricata	Cyperaceae	09.09.09, 15.09.09	
Hänge-Segge	Carex pendula	Cyperaceae	21.08.09, 14.09.09	Drooping sedge
Fuchsrote Segge	Carex buchananii	Cyperaceae	15.09.09	Brown sedge
Gold-Segge	Carex aurea	Cyperaceae	15.09.09	Golden sedge

german name	botanic name	botanic family	date measured	english name
Morgenstern-Segge	Carex Grayii	Cyperaceae	15.09.09	
Hasenfuß-Segge	Carex ovalis	Cyperaceae	15.09.09	Eggbract sedge
Scheinzypergras-Segge	Carex pseudocyperus	Cyperaceae	15.09.09	
Zwerg-Rohrkolben	Typha minima	Typhaceae	11.09.09	Dwarf Bulrush
Breitblättriger Rohrkolben	Typha latifolia	Typhaceae	11.09.09	Common Cattail

Coniferous trees

german name	botanic name	botanic family	date measured	colours	english name
Jeffreays Kiefer	Pinus jeffreyi	Pinaceae	19.10.09		Jeffrey Pine
Korea-Kiefer	Pinus koraiensis	Pinaceae	14.08.09		Korean Pine
Strobe, Weymouths-Kiefer	Pinus strobus	Pinaceae	14.08.09		Eastern White Pine
Rumelische Kiefer	Pinus peuce	Pinaceae	14.08.09		Macedonian Pine
Schlangenhaut-Kiefer	Pinus heldreichii	Pinaceae	14.08.09		Bosnian Pine
Schwarz-Kiefer	Pinus nigra ssp. Nigra	Pinaceae	12.08.09, 14.08.09		European Black Pine
Arve, Zirbelkiefer	Pinus cembra	Pinaceae	14.08.09		Swiss Pine / Arolla Pine
Cilicische Tanne	Abies cilicica	Pinaceae	18.11.09		Syrian Fir Tree

german name	botanic name	botanic family	date measured	colours	english name
Griechische Tanne	Abies cephalonica	Pinaceae	25.07.10		Greek Fir
Numidische Tanne	Abies numidica	Pinaceae	25.07.10		Algerian Fir
Japanische Lärche	Larix kaempferi	Pinaceae	25.07.10		Japanese Larch
Lärche	Larix decidua	Pinaceae	20.08.09		European Larch
Gemeine Lärche	Larix decidua	Pinaceae	27.10.09	x	European Larch
Ginkgo	Ginkgo biloba	Ginkgoaceae	20.10.09	x	Ginkgo / Maidenhair Tree

german name	botanic name	botanic family	date measured	colours	english name
Fichte	Picea	Pinaceae	17.08.09		Norway Spruce
Rauhe Fichte	Picea asperata	Pinaceae	25.07.10		Dragon spruce
Blau-Fichte	Picea pungens 'Glauca'	Pinaceae	25.07.10		Blue spruce
Gemeine Fichte	Picea abies	Pinaceae	25.07.10		Norway spruce
Kaukasische Fichte	Picea orientalis	Pinaceae	25.07.10		Caucasian spruce / Oriental spruce

deciduous trees

german name	botanic name	botanic family	date measured	colours	english name
Purpur-Eukalyptus	Eucalyptus ficifolia (alt) Corymbia ficifolia (neu)	Myrtaceae	18.09.09		Red flowering Gum
	Eucalyptus subcrenulata	Myrtaceae	18.09.09		Tasmanian Alpine Yellow Gum
Gurken-Magnolie	Magnolia acuminata	Magnoliaceae	18.09.09		Cucumber Magnolia
Weidenblättrige Magnolie	Magnolia salicifolia	Magnoliaceae	18.09.09		Cucumber Magnolia
Tulpen-Magnolie	Magnolia x soulangiana	Magnoliaceae	21.08.09, 05.10.09	x	Saucer Magnolia
Kobushi-Magnolie	Magnolia kobus	Magnoliaceae	03.09.09		Kobushi Magnolia
Amerikanischer Tulpenbaum	Liriodendron tulipifera	Magnoliaceae	15.07.09, 21.10.09	x	Tuliptree

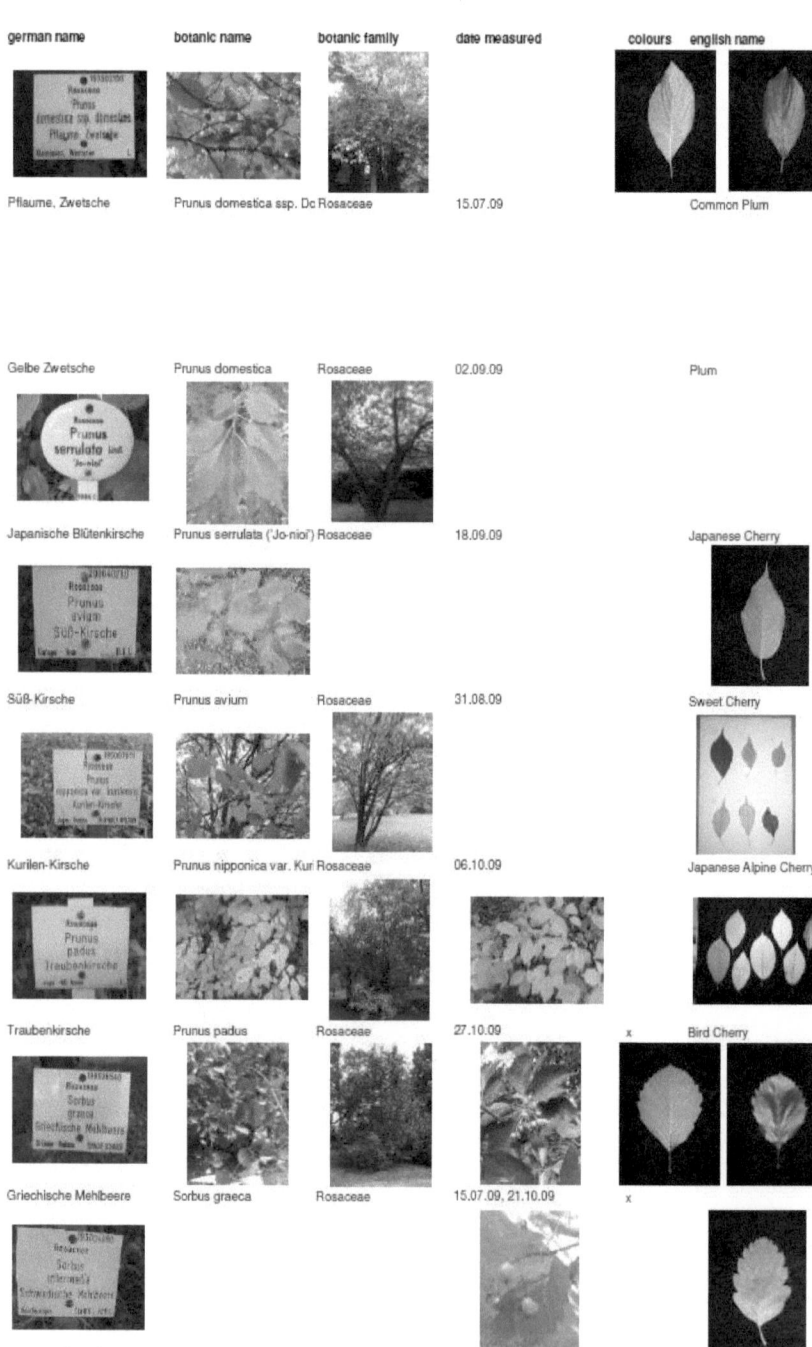

german name	botanic name	botanic family	date measured	colours	english name
Kanadische Felsenbirne	Amelanchier canadensis	Rosaceae	02.09.09		Canadian serviceberry
Schwarzfrüchtiger Weißdorn	Crataegus nigra	Rosaceae	02.09.09		
Fiederblattweißdorn	Crataegus pinnatifida	Rosaceae	03.09.09		Chinese Hawthorn
Silber-Pappel	Populus alba	Salicaceae	14.07.09, 21.10.09	x	White Poplar
Sibirische Ulme	Ulmus pumila	Ulmaceae	31.08.09		Siberian Elm
Schönulme	Euptelea pleiosperma	Eupteleaceae	15.07.09		Pleiospermum
Silber-Linde	Tilia tomentosa	Tiliaceae	18.09.09		Silver Lime (UK) Silver Linden (USA)
Winter-Linde	Tilia cordata	Tiliaceae	27.04.09, 20.08.09, 26.08.09, 31.08.09, 07.10.09	x, x	Small-leaved Lime (UK) Small-leaved Linden (USA)

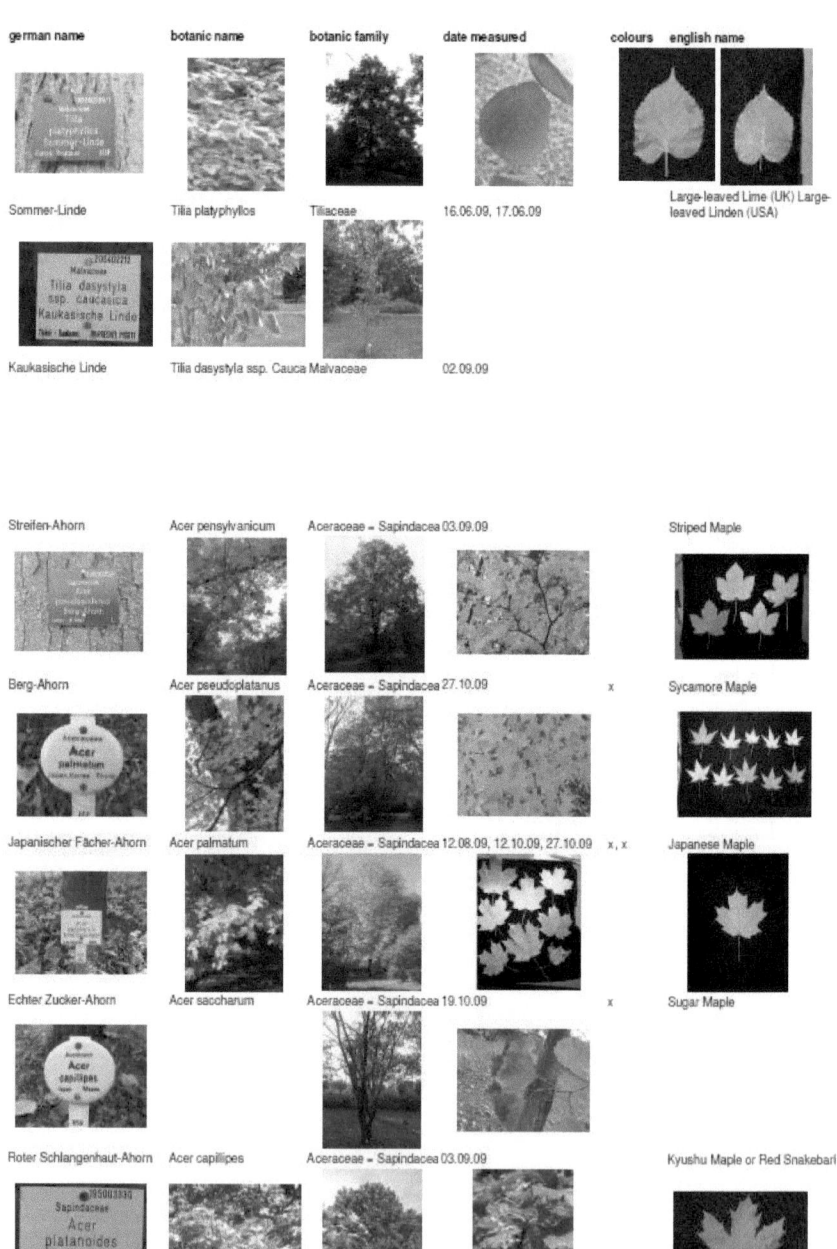

german name	botanic name	botanic family	date measured	colours	english name
Sommer-Linde	Tilia platyphyllos	Tiliaceae	16.06.09, 17.06.09		Large-leaved Lime (UK) Large-leaved Linden (USA)
Kaukasische Linde	Tilia dasystyla ssp. Cauca	Malvaceae	02.09.09		
Streifen-Ahorn	Acer pensylvanicum	Aceraceae – Sapindacea 03.09.09			Striped Maple
Berg-Ahorn	Acer pseudoplatanus	Aceraceae – Sapindacea 27.10.09		x	Sycamore Maple
Japanischer Fächer-Ahorn	Acer palmatum	Aceraceae – Sapindacea 12.08.09, 12.10.09, 27.10.09		x, x	Japanese Maple
Echter Zucker-Ahorn	Acer saccharum	Aceraceae – Sapindacea 19.10.09		x	Sugar Maple
Roter Schlangenhaut-Ahorn	Acer capillipes	Aceraceae – Sapindacea 03.09.09			Kyushu Maple or Red Snakebark
Spitz-Ahorn	Acer platanoides	Aceraceae – Sapindacea 16.06.09, 17.06.09			Norway Maple

german name	botanic name	botanic family	date measured	colours	english name
Japanische Zaubernuss	Hamamelis japonica	Hamamelidaceae	12.10.09	x	Japanese Snapping Hazel
Orientalischer Amberbaum	Liquidambar orientalis	Hamamelidaceae	18.11.09	x	Oriental Sweetgum
Taschentuchbaum	Davidia involucreta var. Vi	Davidiaceae	14.07.09		Dove Tree
Mongolischer Maulbeerbaum	Morus mongolica	Moraceae	31.08.09, 07.10.09	x	Mongolian Mulberry
Chinesischer Maulbeerbaum	Morus cathayana	Moraceae	03.09.09		
Osagedorn	Maclura pomifera	Moraceae	03.09.09		Osage-Orange
Alexandrischer Lorbeer	Danae racemosa	Ruscaceae	04.11.09	x	Alexandrian Laurel
Japanische Walnuss	Juglans ailantifolia	Juglandaceae	06.10.09	x	Japanese Walnut

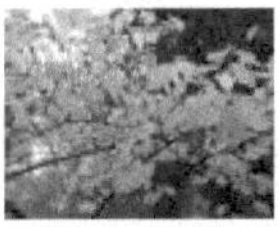

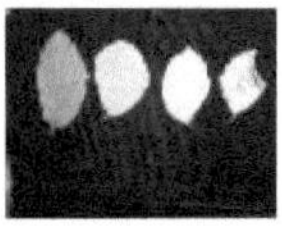

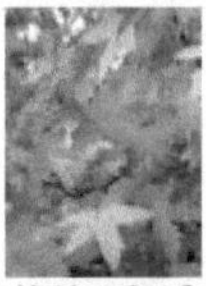

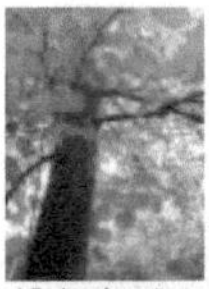

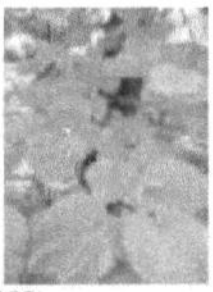

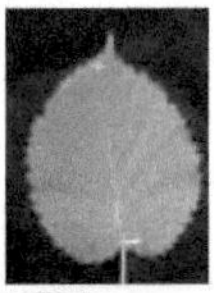

german name	botanic name	botanic family	date measured	colours	english name
Japanischer Schnurbaum	Sophora japonica	Leguminosae	13.11.09	x	Pagoda Tree
Baumaralie	Kalopanax septemlobus	Araliaceae	06.10.09	x	Kalopanax / Prickly Castor-oil Tree
Birke	Betula spec.	Betulaceae	14.07.09, 03.09.09, 06.10.09	x	Birch
Lindenblättrige Birke	Betula maximowicziana	Betulaceae	03.09.09		
Haselnuss	Corylus avellana	Betulaceae	21.10.09	x	Cobnut / Hazel
Kaukasische Erle	Alnus subcordata	Betulaceae	02.09.09		Caucasian Alder
Grau-Erle	Alnus incana	Betulaceae	15.07.09		Grey Alder
Schwarz-Erle	Alnus glutinosa	Betulaceae	02.09.09		Black Alder

german name	botanic name	botanic family	date measured	colours	english name
Gewöhnliche Hainbuche	Carpinus betulus	Betulaceae	12.08.09, 14.08.09, 20.08.09, 31.08.09		Hornbeam
Götterbaum	Ailanthus altissima	Simaroubaceae	20.08.09		Ailanthus / Tree of Heaven

deciduous others

german name	botanic name	botanic family	Picture	date measured	english name
Laubholz-Mistel	Viscum album ssp. album	Santalaceae		07.10.09	Common Mistletoe
Filzige Pfeifenwinde	Aristolochia tomentosa	Aristolochiaceae		14.07.09	Dutchman's Pipevine
Eselsohr / Woll-Ziest	Stachys byzantina	Labiatae		14.07.09	Lamb's Ear
Kaukasus-Efeu	Hedera colchica	Araliaceae		04.09.09	Persian Ivy

Appendix B: effect of changing illumination and viewing angle on different leaves

This appendix shows the results of the measurements performed to evaluate the effects of changing illumination and viewing angles on reflectance from leaves. This is discussed in Chapter 8.7.

Acer platanoides

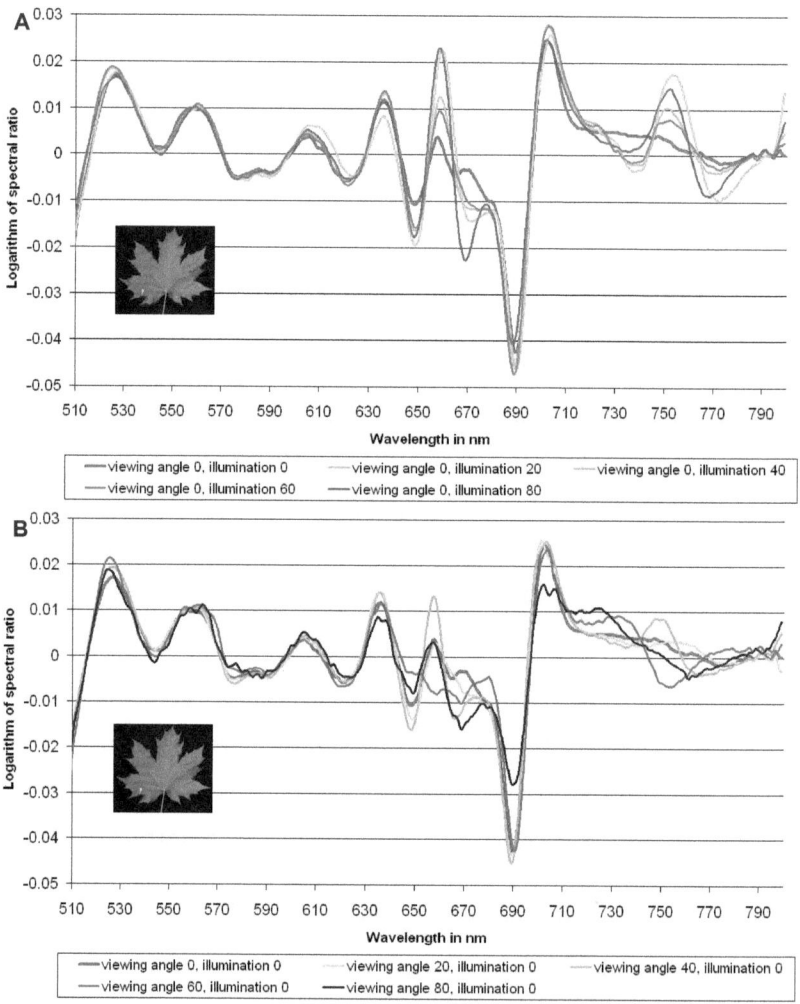

Figure B.1
Effect of changing illumination and viewing angle on high-frequency reflection structures of a maple leaf.
A) shows the effect of changing illumination angle. B) shows the effect of changing viewing angle.

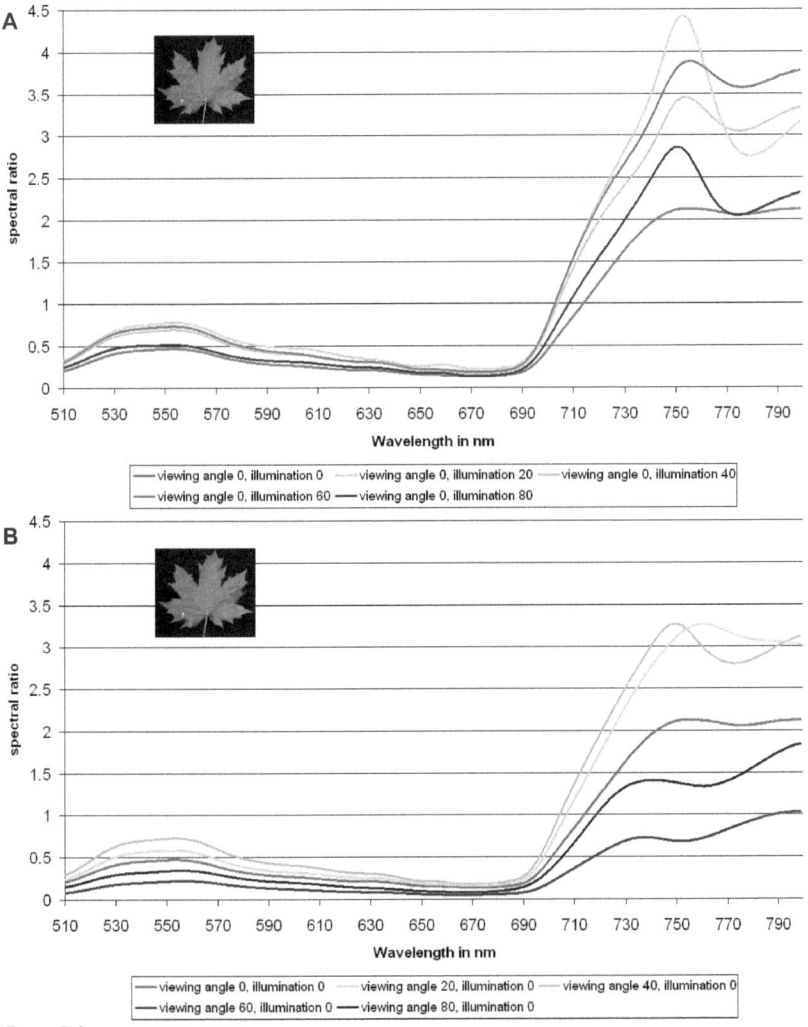

Figure B.2
Effect of changing illumination and viewing angle on soectral ratio of reflection of a maple leaf. A) shows the effect of changing illumination angle. B) shows the effect of changing viewing angle.

Prunus domestica

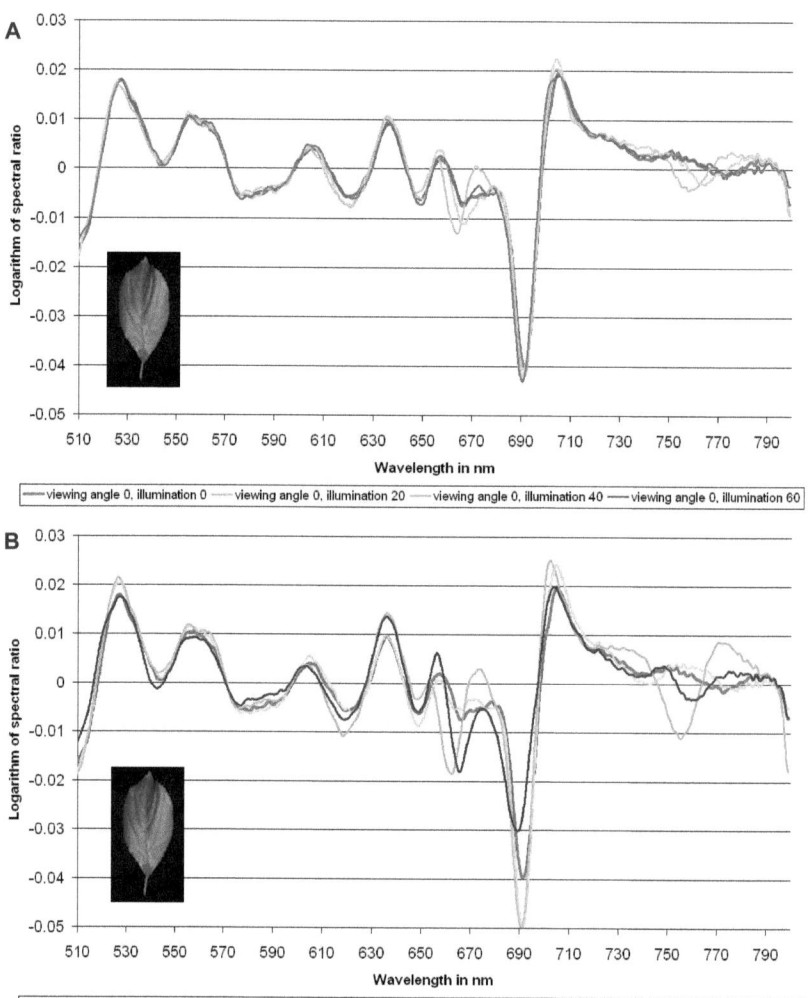

Figure B.3
Effect of changing illumination and viewing angle on reflectance of a plum leaf. A) shows the effect of changing illumination angle. B) shows the effect of changing viewing angle.

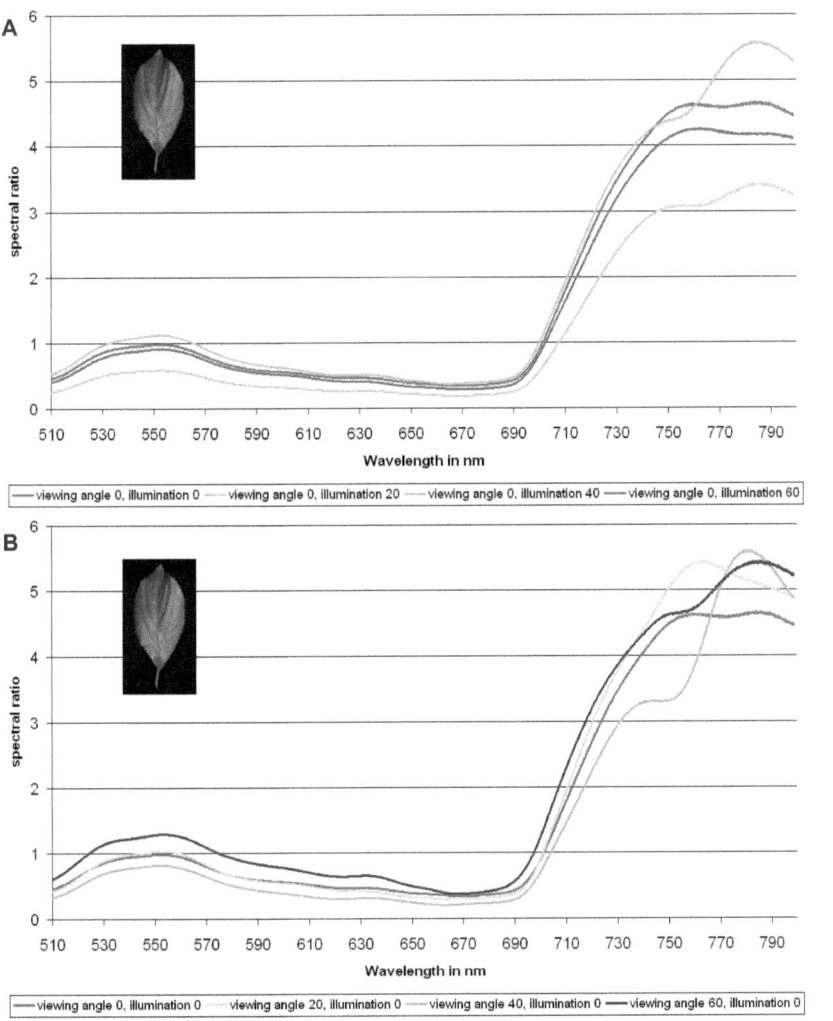

Figure B.4
Effect of changing illumination and viewing angle on soectral ratio of reflection of a plum leaf. A) shows the effect of changing illumination angle. B) shows the effect of changing viewing angle.

Liriodendron Tulipifera

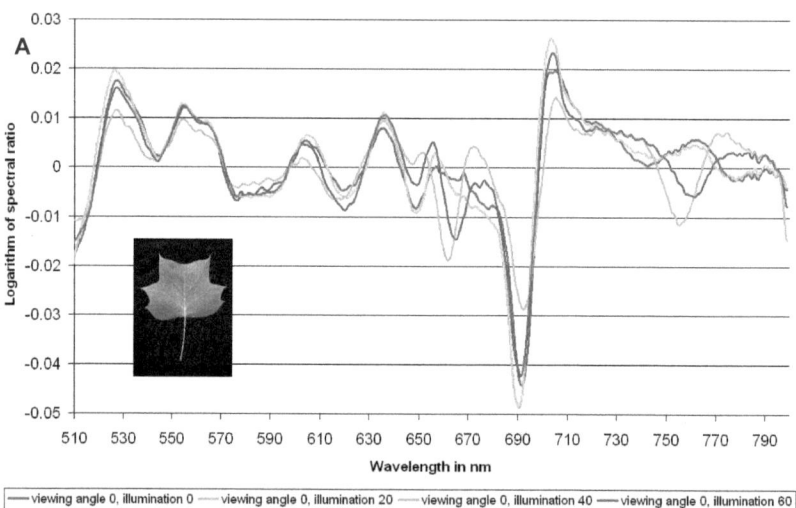

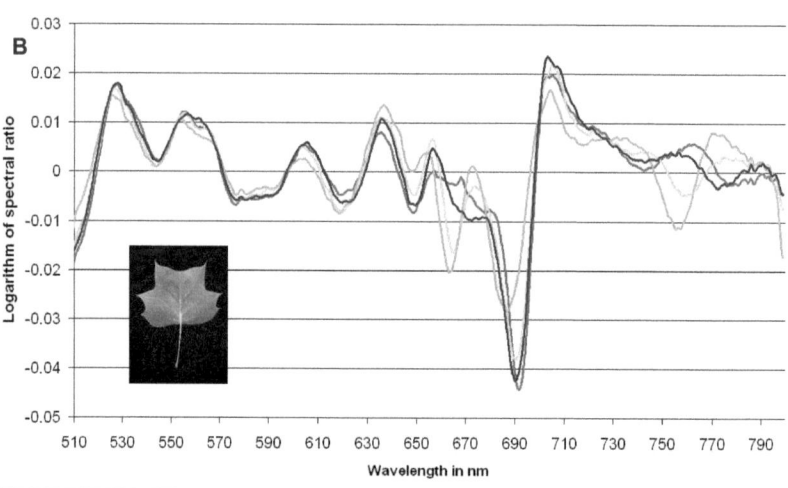

Figure B.5
Effect of changing illumination and viewing angle on reflectance of a Tulip tree leaf. A) shows the effect of changing illumination angle. B) shows the effect of changing viewing angle.

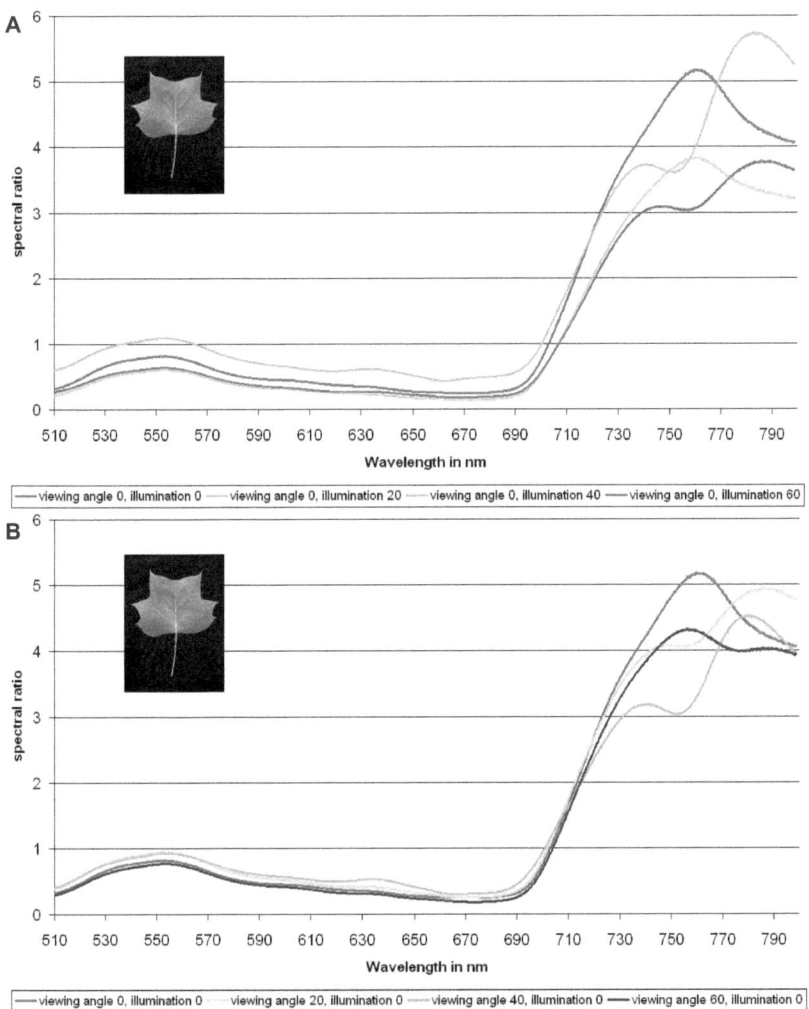

Figure B.6
Effect of changing illumination and viewing angle on soectral ratio of reflection of a Tulip tree leaf. A) shows the effect of changing illumination angle. B) shows the effect of changing viewing angle.

Euptelea pleiospermum

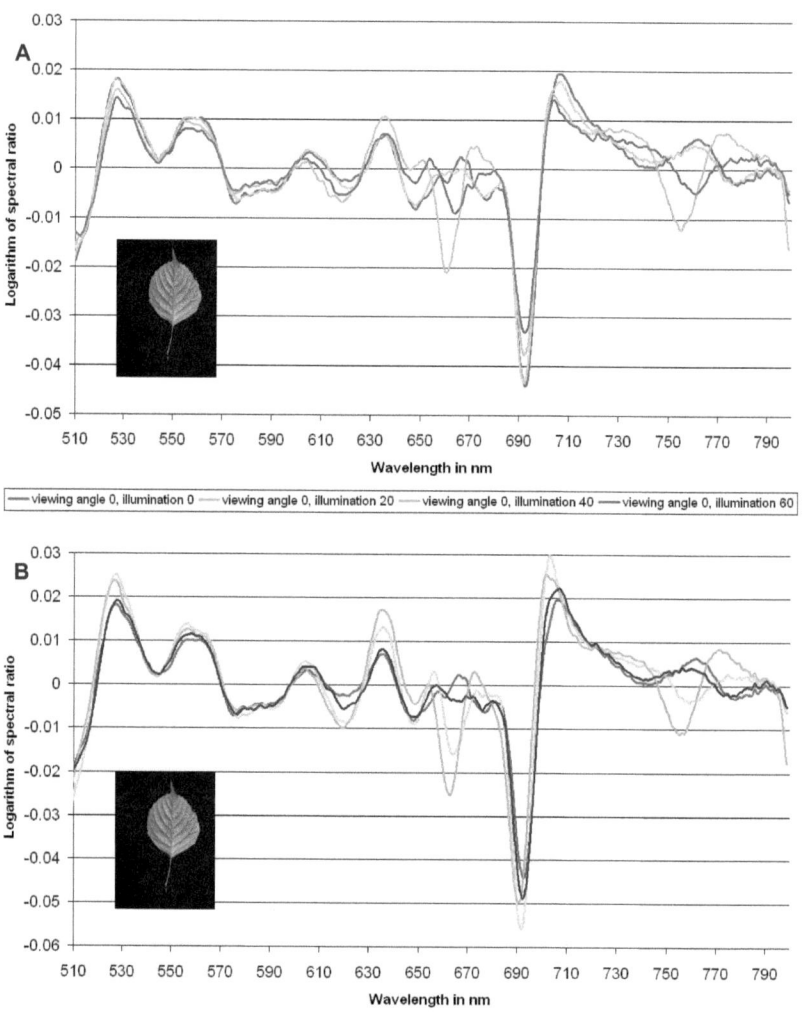

Figure B.7
Effect of changing illumination and viewing angle on reflectance of an Euptelea pleiospermum leaf. A) shows the effect of changing illumination angle. B) shows the effect of changing viewing angle.

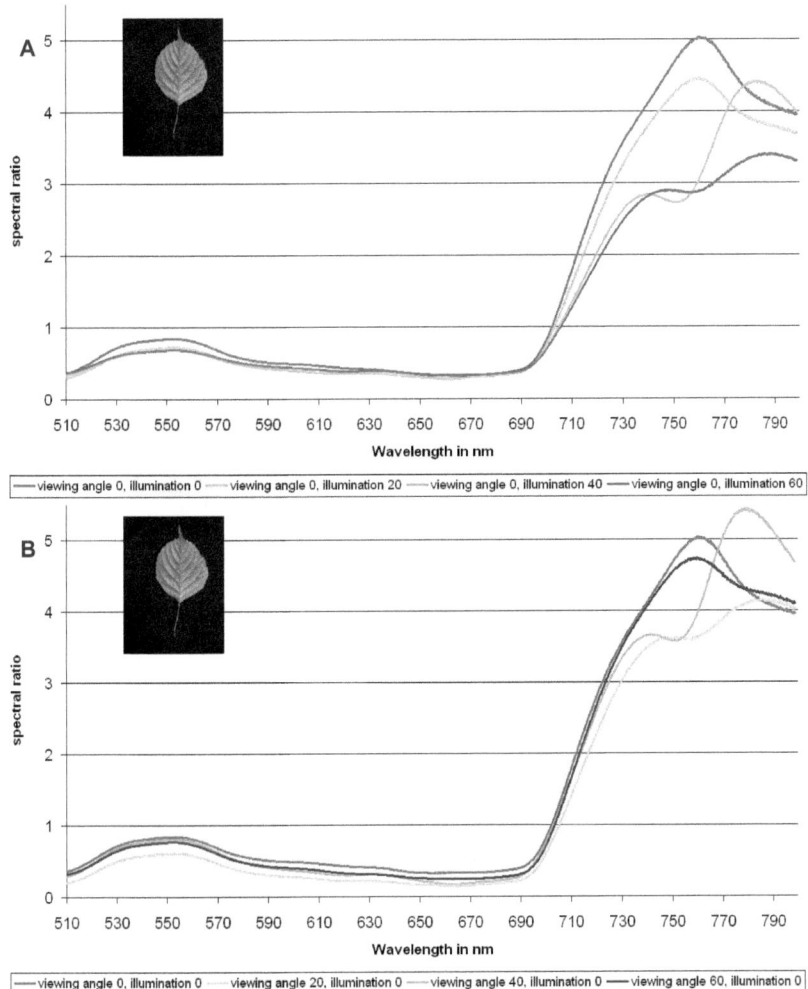

Figure B.8
Effect of changing illumination and viewing angle on soectral ratio of reflection of a Euptelea pleiospermum leaf. A) shows the effect of changing illumination angle. B) shows the effect of changing

Viburnum rhytidophyllum

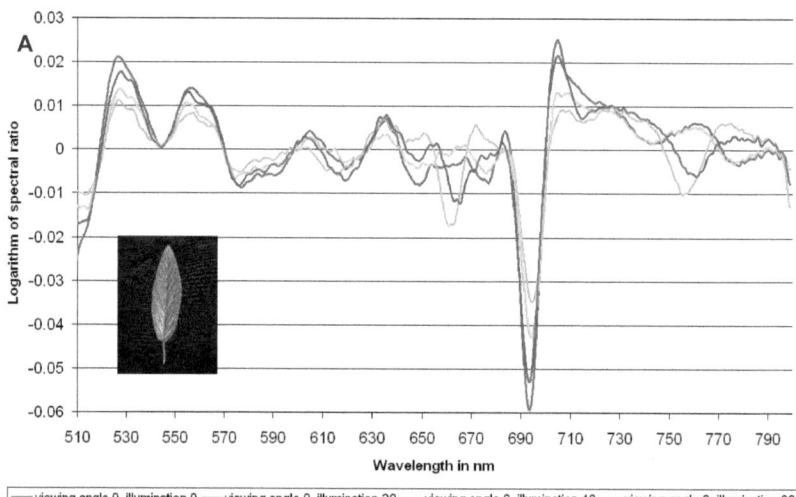

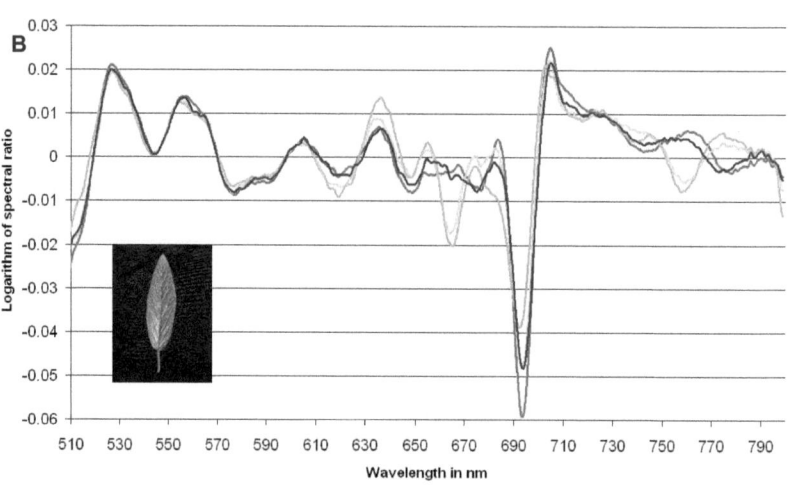

Figure B.9
Effect of changing illumination and viewing angle on reflectance of a Leatherleaf Viburnum leaf. A) shows the effect of changing illumination angle. B) shows the effect of changing viewing angle.

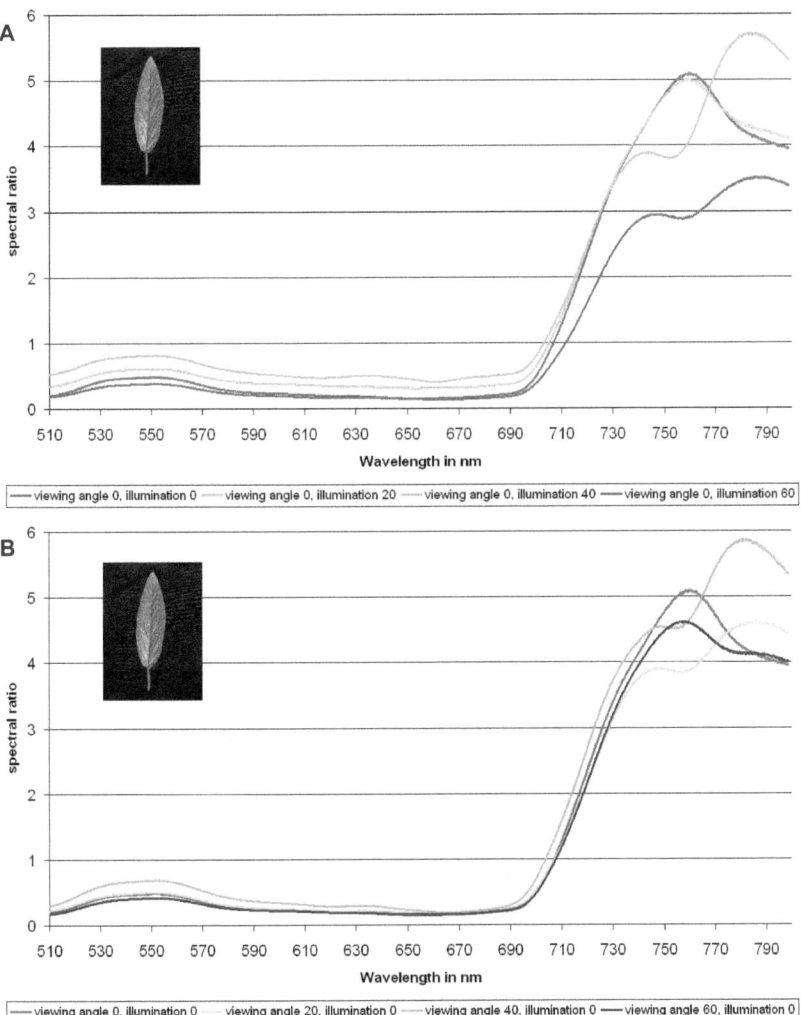

Figure B.10
Effect of changing illumination and viewing angle on soectral ratio of reflection of a Leatherleaf Viburnum leaf. A) shows the effect of changing illumination angle. B) shows the effect of changing viewing angle.

Alnus incana

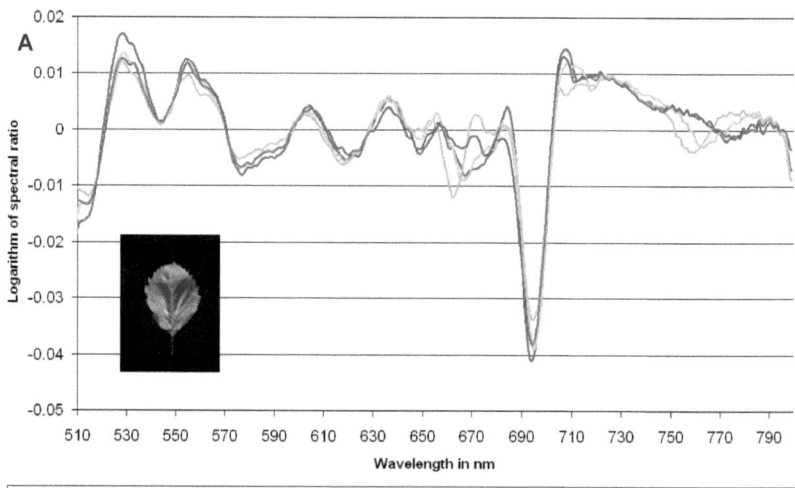

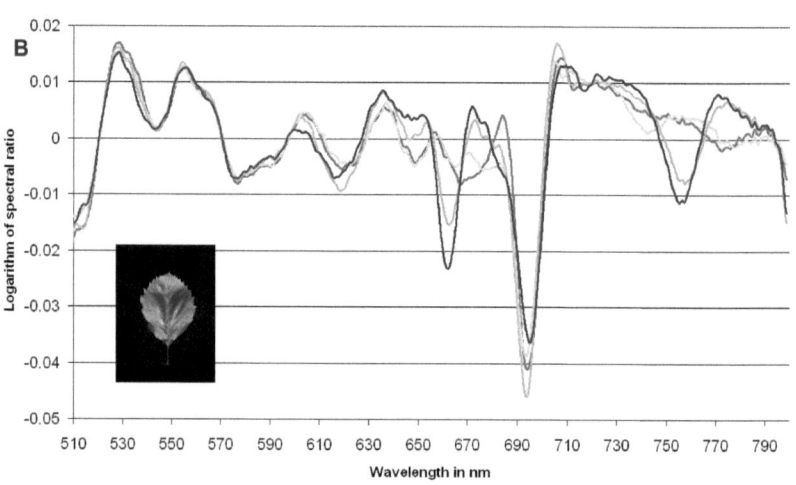

Figure B.11
Effect of changing illumination and viewing angle on reflectance of a Grey Alder leaf. A) shows the effect of changing illumination angle. B) shows the effect of changing viewing angle.

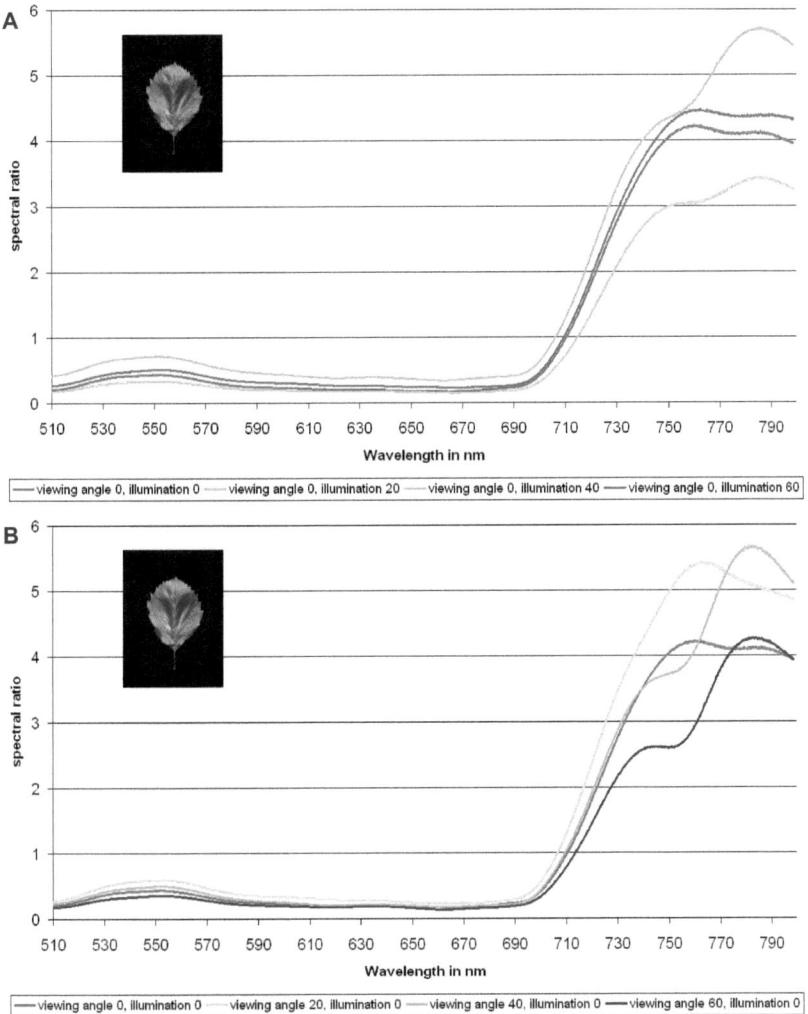

Figure B.12
Effect of changing illumination and viewing angle on soectral ratio of reflection of a Grey Alder leaf. A) shows the effect of changing illumination angle. B) shows the effect of changing viewing angle.

Sorbus graeca

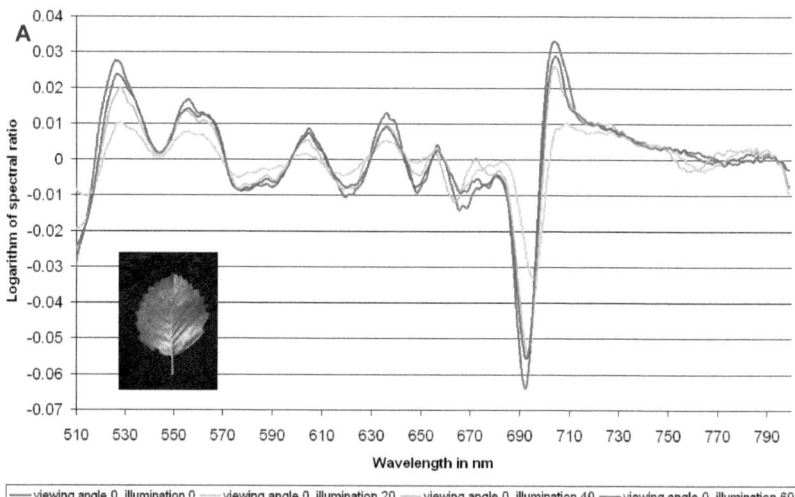

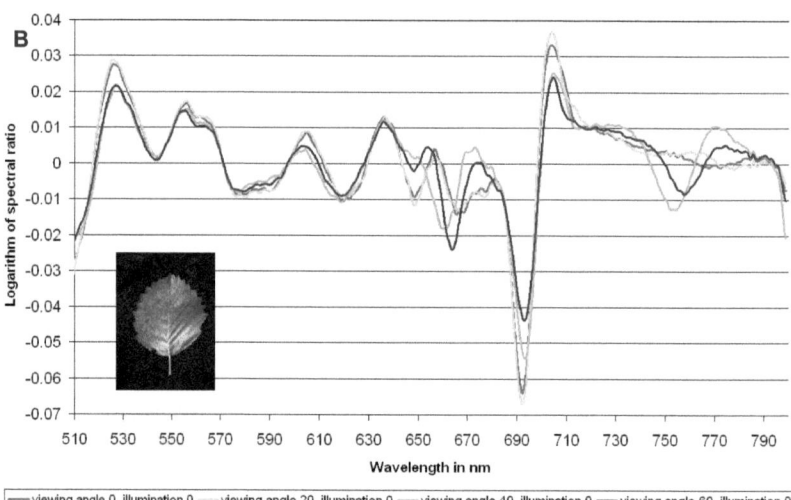

Figure B.13
Effect of changing illumination and viewing angle on reflectance of a Sorbus graeca leaf. A) shows the effect of changing illumination angle. B) shows the effect of changing viewing angle.

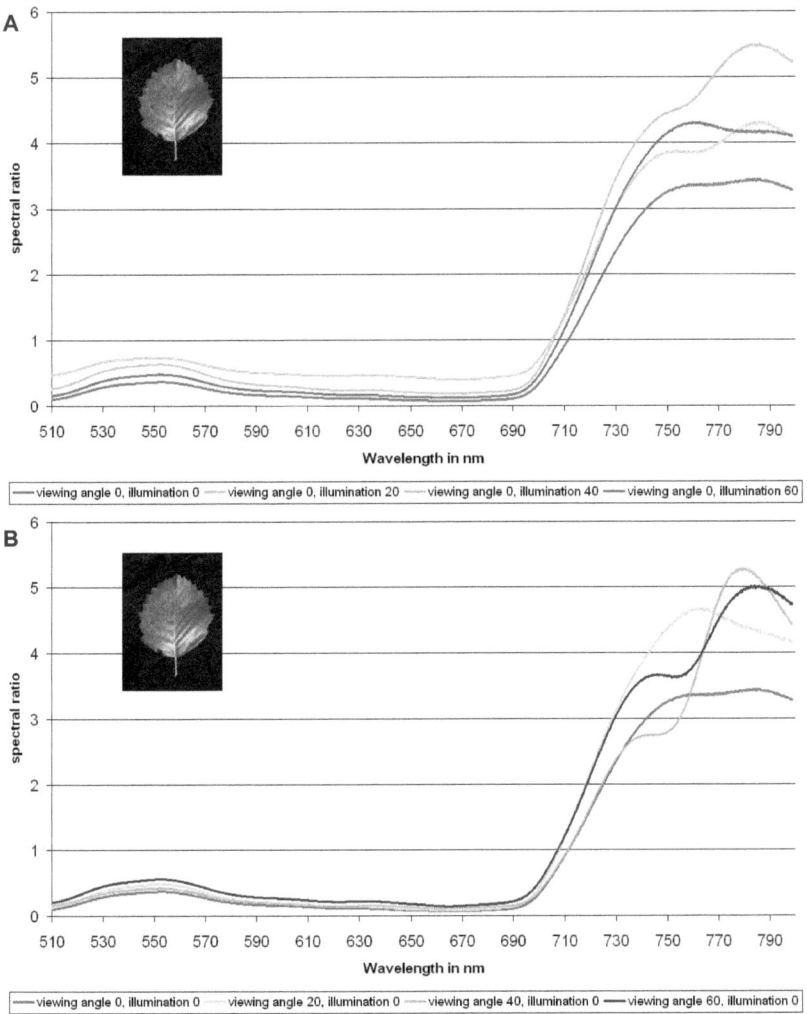

Figure B.14
Effect of changing illumination and viewing angle on soectral ratio of reflection of a Sorbus graeca leaf. A) shows the effect of changing illumination angle. B) shows the effect of changing viewing angle.

Davidia involucreta

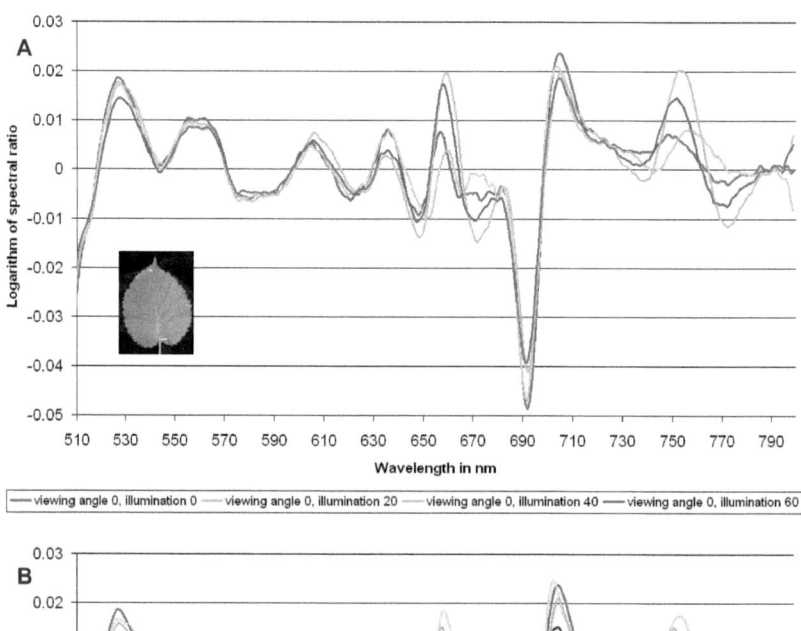

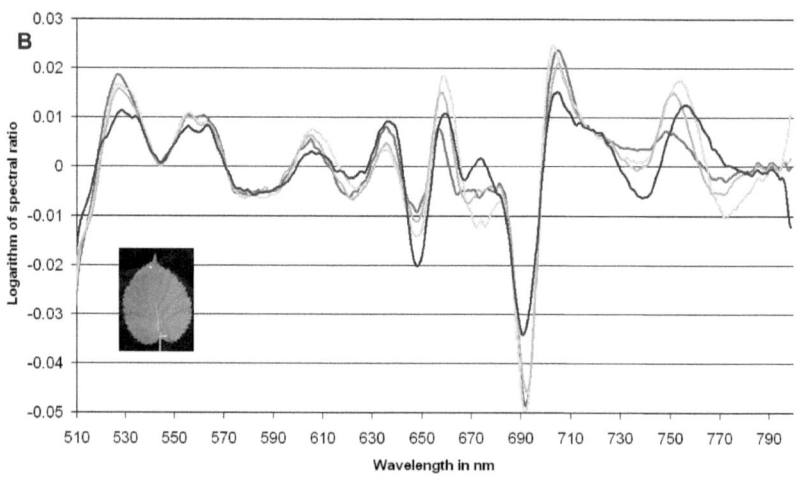

Figure B.15
Effect of changing illumination and viewing angle on reflectance of a Dove Tree leaf. A) shows the effect of changing illumination angle. B) shows the effect of changing viewing angle.

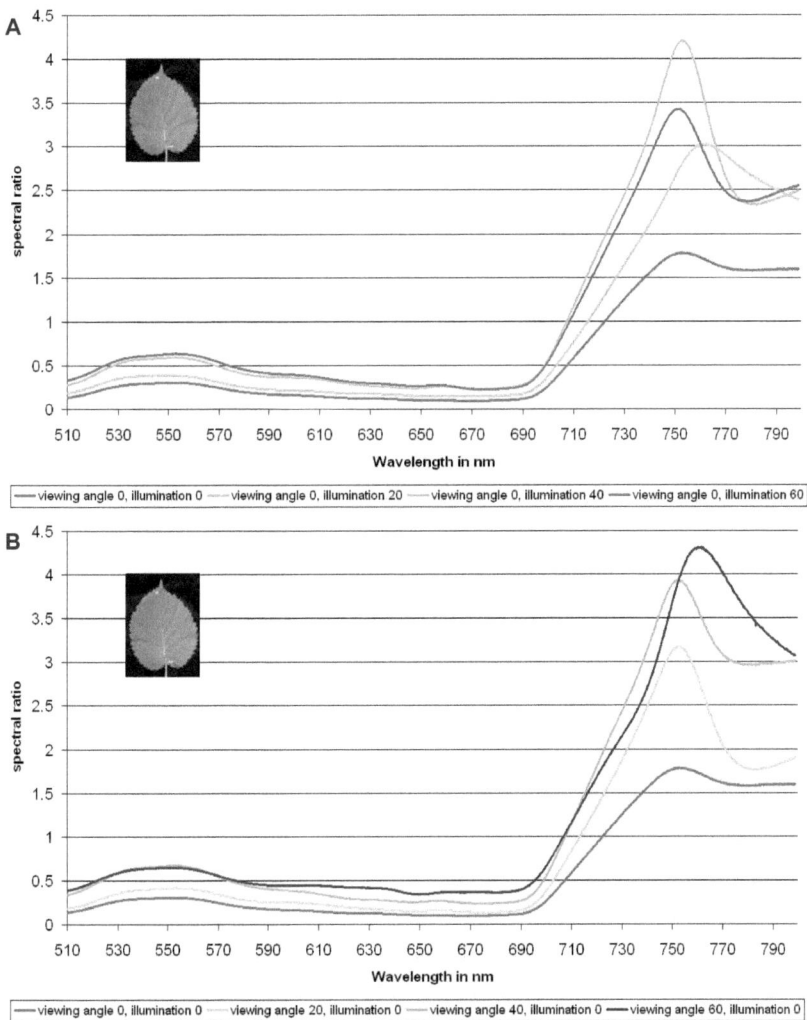

Figure B.16
Effect of changing illumination and viewing angle on soectral ratio of reflection of a Dove Tree leaf. A) shows the effect of changing illumination angle. B) shows the effect of changing viewing angle.

Zea mays

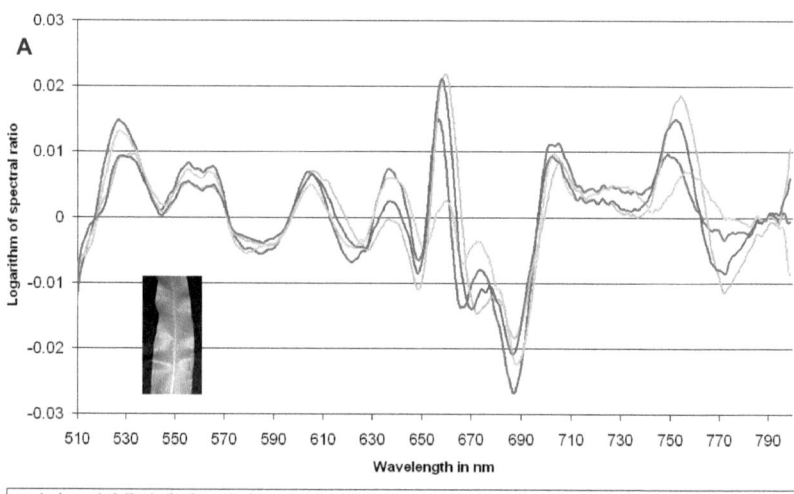

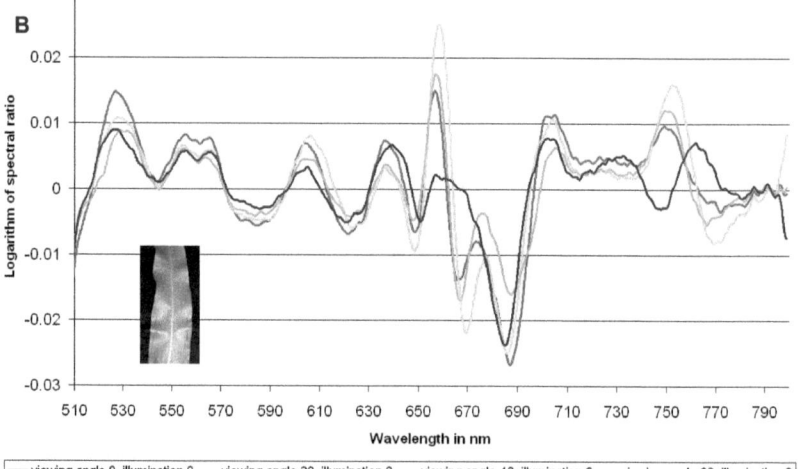

Figure B.17
Effect of changing illumination and viewing angle on reflectance of a Maize leaf. A) shows the effect of changing illumination angle. B) shows the effect of changing viewing angle.

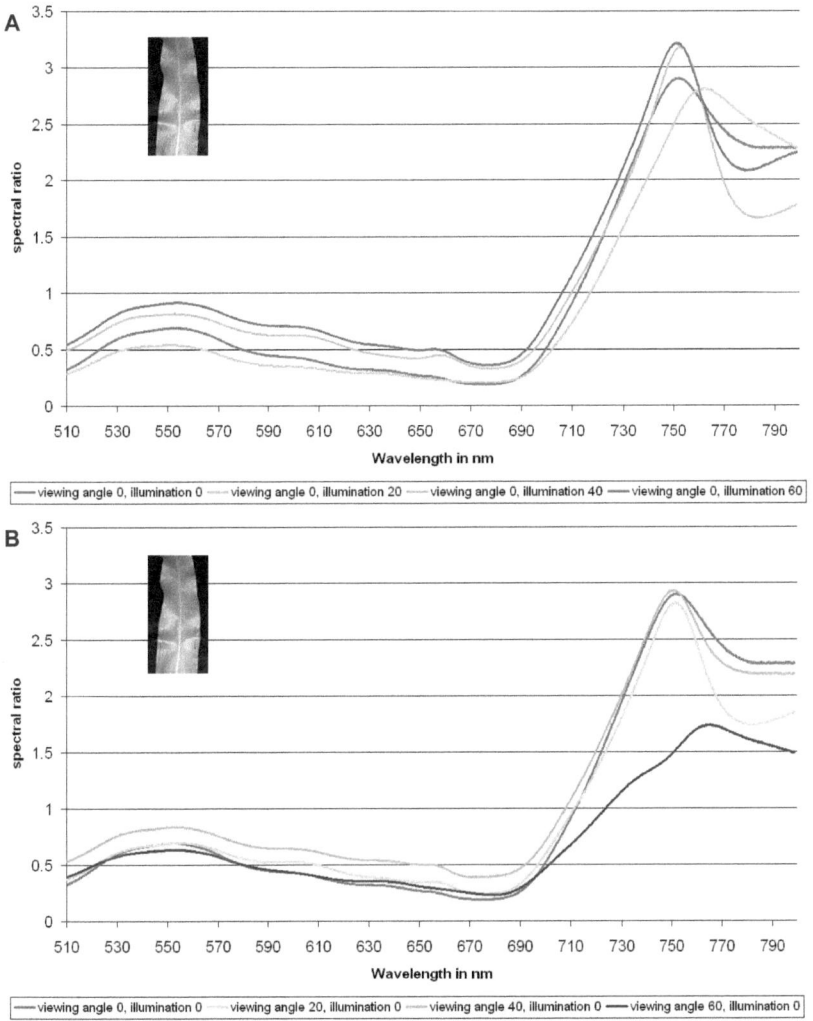

Figure B.18
Effect of changing illumination and viewing angle on soectral ratio of reflection of a Maize leaf. A) shows the effect of changing illumination angle. B) shows the effect of changing viewing angle.

Stachys byzantina

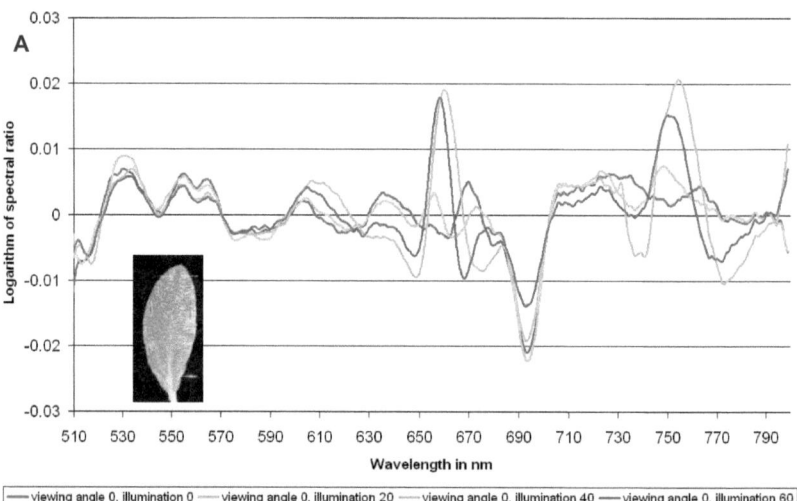

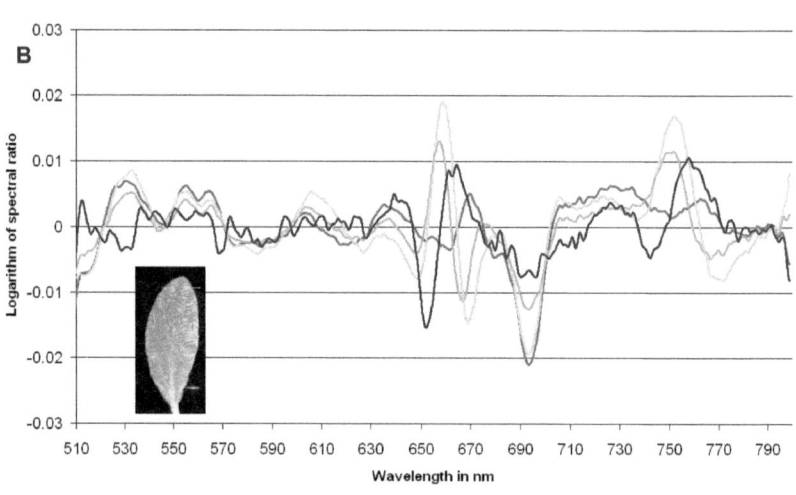

Figure B.19
Effect of changing illumination and viewing angle on reflectance of a Lamb's Ear leaf. A) shows the effect of changing illumination angle. B) shows the effect of changing viewing angle.

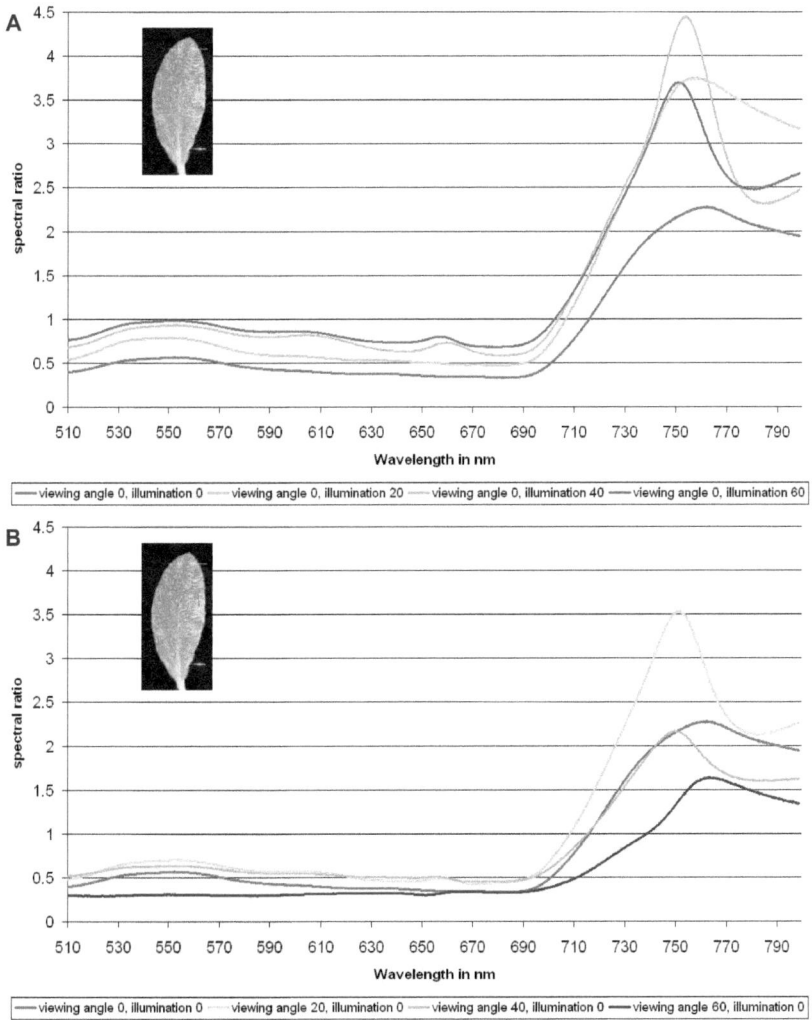

Figure B.20
Effect of changing illumination and viewing angle on soectral ratio of reflection of a Lamb's Ear leaf. A) shows the effect of changing illumination angle. B) shows the effect of changing viewing angle.

I want morebooks!

Buy your books fast and straightforward online - at one of world's fastest growing online book stores! Environmentally sound due to Print-on-Demand technologies.

Buy your books online at
www.morebooks.shop

Kaufen Sie Ihre Bücher schnell und unkompliziert online – auf einer der am schnellsten wachsenden Buchhandelsplattformen weltweit! Dank Print-On-Demand umwelt- und ressourcenschonend produziert.

Bücher schneller online kaufen
www.morebooks.shop

KS OmniScriptum Publishing
Brivibas gatve 197
LV-1039 Riga, Latvia
Telefax: +371 686 204 55

info@omniscriptum.com
www.omniscriptum.com

Printed by Books on Demand GmbH, Norderstedt / Germany